C000144107

British Life in India

There are good things, there are some indifferent, there are more that are bad that you read here. Not otherwise, Avitus, is a book produced.

Martial

British Life in India

An Anthology of Humorous and Other Writings
Perpetrated by the British in India 1750–1947
with Some Latitude for Works
Completed after Independence

Edited by

R.V. Vernède

Delhi
Oxford University Press
Bombay   Calcutta   Madras
1995

Oxford University Press, Walton Street, Oxford OX2 6DP

Oxford New York Toronto
Delhi Bombay Calcutta Madras Karachi
Kuala Lumpur Singapore Hong Kong Tokyo
Nairobi Dar es Salaam Cape Town
Melbourne Auckland Madrid

and associates in
Berlin Ibadan

ISBN 0 19 563486 1

Text illustrations by Dean Gasper
Jacket illustration by Meera Dayal Deshaprabhu

Typeset by Guru Typograph Technology, New Delhi 110045
printed at Pauls Press, New Delhi 110020
and published by Neil O'Brien, Oxford University Press
YMCA Library Building, Jai Singh Road, New Delhi 110001

The Cummerbund[1]—An Indian Poem

I

She sate upon her Dobie,[2]
to watch the Evening Star,
and all the Punkahs[3] as they passed
cried, 'My! how fair you are!'
Around her bower, with quivering leaves,
the tall khansamahs[4] grew,
and khitmutgars[5] in wild festoons
hung down from tchokis[6] blue.

II

Below her home the river rolled
with soft meloobious sound,
where golden-finned chuprassies[7] swam,
in myriads circling round.
Above, on tallest trees remote,
green ayahs[8] perched alone,
and all night long the mussak[9] moan'd
its melancholy tone.

III

And where the purple nullahs[10] threw
their branches far and wide,—
and silvery goreewallahs[11] flew
in silence, side by side,—

---

[1] A silk waist sash, mostly black, worn with white evening dress. A smarter alternative to braces. [2] washerman [3] large swinging cloth fan on wooden frame, pulled by a cord or leather thong [4] cooks [5] waiters [6] police stations or lock-ups [7] orderlies [8] English lady's personal maid and nanny for children [9] goatskin bag in which water was carried [10] ditches, dry or wet watercourses [11] grooms.

the little bheesties'[12] twittering cry
rose in the fragrant air,
and oft the angry Jampan[13] howled
deep in his hateful lair.

IV

She state upon her Dobie,—
she heard the Nummak[14] hum,—
when all at once a cry arose:
'The Cummerbund is come!'
In vain she fled;—with open jaws
the angry monster followed,
and so (before assistance came),
that Lady Fair was swollowed.

V

They sought in vain for even a bone
respectfully to bury,—
they said, 'Hers was a dreadful fate!'
(and Echo answered, 'Very.')
They nailed her Dobie to the wall,
where last her form was seen,
and underneath they wrote these words,
in yellow, blue, and green:

VI

'Beware, ye Fair! Ye Fair, Beware!
nor sit out late at night,—
lest horrid Cummerbunds should come,
and sollow you outright.'

Edward Lear
First published in the *Times of India*,
Bombay, July 1874
Included in his *Indian Journal*, 1873–1875

---

[12] Water-carriers   [13] a kind of sedan chair in great demand in hill-stations where no cars were allowed   [14] salt.

Contents

1
People

NEWCOMERS

EDITOR'S INTRODUCTION

From about 1780, three categories of fairly regular newcomers to India can clearly be distinguished. Though the numbers of newcomers outside these categories did gradually increase, it was not until the opening of the Suez Canal in 1869 that the tourists began to outnumber the Civil Servants and the soldiers travelling to India. The three categories were:

1. The young men recruited by the East India Company, originally by overt patronage, to fill the posts of writer and later to become collectors of revenue, judges and magistrates in the areas where the Company was effectively in control. They had no previous training for these posts. In 1856 some of these young men were recruited by competitive examination and from 1858 onwards all were recruited to various Imperial Services by competition, but now as servants of the Crown.

2. Other young men continued to be recruited by patronage to join business houses, factories, mills, and indigo, jute, tea and coffee plantations. They were not so well educated as the 'competition-*wallahs*', though many of them went to public schools. Some of them became successful businessmen and some of their firms became household words in the East.

3. From at least 1780, and probably from earlier, a handful of daring young women, braving the hazards and hardships of the journey from England, sailed to India every year, their main or

---

Competition-*wallah* someone who competes for a place in the ICS by sitting for the competitive exam, not by patronage

only objective being to hook a wealthy husband. No doubt their ambition owed something to the stories they heard in England, about the wealth and ostentation of the Nabobs. They came to be known as 'the fishing fleet' and they made for the larger places by the sea, Calcutta, Bombay, Madras and Rangoon. Their determination and persistence were to be admired (or feared?), for this traffic continued from the eighteenth century up to the outbreak of the Second World War. For some it was a desperate gamble, for they had not the funds to return to England. These young women, if they failed to catch a wealthy husband, took posts as governesses or nannies with some British family. After some years, some at least would find more suitable, though poorer, men to marry. Girls who failed to find wealthy husbands but who could afford to go home, were referred to unkindly as 'returned empties'.

Up to about 1850, the voyage to India was made in sailing ships. The passage round the Cape was slow, uncomfortable and sometimes dangerous. In addition to hazard by storm, during the whole period when Britain and France were at war in Europe, hostilities spilt over into the Indian Ocean and there was the added hazard of being captured by a French frigate. If trade winds were favourable the voyage might take four months, if unfavourable, six months. The old sailing ships were not designed to carry passengers, especially female passengers. They were accommodated either in the 'Round House' or in the 'Great Cabin' originally intended for the ship's officers. Knowledgeable travellers took their own livestock aboard—hens, ducks, sheep, etc.—to supplement or vary the ship's diet. Some time early in the nineteenth century the ships' owners fitted cabins—18 was about the usual number—but the passengers had to furnish their own cabins.

In the old days of sail with only a very limited number of passengers, a crowd of curious onlookers went down to the docks to watch the passengers disembark from the latest arrived ship. If there was a pretty girl, you may be sure that tongues wagged busily until her identity, age, and the names of her host and hostess or relations were established. In Madras there was the added excitement of watching the passengers come ashore through three lines of dangerous surf in a *Masala* boat, a flat-bottomed boat built without any iron, the planks being sewn together with line made from the outer husks of coconuts.

Being lighter and more flexible than a conventional boat, the Masala could ride on top of the surf, but the boatmen had to be highly skilled. As the last line of surf carried the boat shoreward, the boatmen leapt out and carried their passengers on their shoulders to the beach. One or two smaller boats accompanied each Masala boat, to rescue its passengers in case the Masala should overturn and spill its occupants into the sea.

When the novelty of steamships had worn off no one bothered to meet the ships, except those who were meeting a passenger.

Steamships were introduced on the run to India in the 1850s. At first, like sailing ships, they went round the Cape, but very soon what was known as 'the overland' route came into use. Passengers travelled by sea from London or Marseilles to Alexandria and then by train to just beyond Cairo, as far as the line had been built. From this point they travelled in pony-drawn minibuses to Suez where they embarked on another steamer for India. This journey took from four to six weeks as compared to four to six months by sail. A number of wives took advantage of this improvement to join their husbands in India and some of them also took their daughters with them. I suggest this may account, to some extent, for the surprisingly large number of British women found to be in India in 1857.

After the Suez Canal was opened in 1869, the flow of visitors to India became a steady trickle, if not yet a flood, and was conditioned by the inadequate provision of hotels and amenities. As a result of these restrictions, visitors who had relatives or friends in India went out to stay with them. Others, very largely drawn from the ranks of the British aristocracy, more or less invited themselves to stay with the Viceroy, the provincial governors and occasionally with other senior members of Government. Their presence was not always welcome. Another phenomenon, not previously known, was British Members of Parliament who came out ostensibly to improve their knowledge of India, but whose ignorance sometimes made them a laughing stock amongst 'old India hands'.

In the days of sail and before the construction of the splendid Indian Railway Systems, arrival at, say, Calcutta was by no means the end of every journey from England. Let us suppose that the final destination of our newcomer was Allahabad or Lucknow, a slow and uncomfortable journey of 500 miles by road to Allahabad, and 670 miles to Lucknow, was necessary. There were three methods of inland travel:

1. By pony and *dawk* trap, with daily stages normally of ten or twelve miles, the distance between dawk bungalows where the traveller would spend the night. It was useless to hope you might do a double stage, because the ponies used in the trap were so weak and half-starved that they could only complete ten or twelve miles with the greatest difficulty and the next day you would have a change of pony.

2. By *palki* or palanquin. This was the only acceptable alternative if the road was too rough for a vehicle. In the south of India they boasted that their palki service was as fast as the pony dawk of the north. There were four carriers to a palki with a relief team following behind. Carriers went at a fast walk. With a very heavy passenger they had to go slower but expected a higher wage. They might also sing a lively and unflattering refrain about the weight of their fare.

3. By river, in native boats *budgerows*. This was the slowest, although all heavy baggage was sent by boat where possible. The boat was more comfortable than the pony trap and the palki, but this was more than offset by the greater distance and the greater time taken for the journey, not to mention the constant delays caused by running on to shifting sandbanks. The distance by the river Ganges from Calcutta to Allahabad was 800 miles, which would take about three months by boat. When river steamers appeared on the Ganges the speed of this method of travel was greatly improved, but, of course, could not be compared to the sixteen days taken by Captain William Peel to transport the naval brigade by river steamer from Calcutta to Allahabad in 1857. Travel by river steamer never caught on because the railways, coming hard on its heels stole all the goods and passenger traffic.

Arrived at last in the station to which he had been assigned, the newcomer faced his final obstacle—the business of 'calling'. The practice has been ridiculed and was not really necessary in a really small station. But in a large station, despite the element of snobbery which everywhere accompanies etiquette, it did serve the useful

---

Dawk mode of transportation by relay of men and horses   *dawk bungalows* rest house for travellers   *palki* palanquin   *budgerows* country river boats

purpose of enabling the station to take note of a new arrival who might otherwise be missed, and of noting that an old resident who might have been away for some time was back and still alive.

A newcomer to India was known as a 'griffin'. This term could be applied to new recruits to all services, but the Indian Civil Service, with their usual arrogance, considered that the designation applied exclusively to their own young.

*

The Fishing Fleet

The Madras correspondent of Hicky's Bengal Gazette reported, in 1781, that a bevy of eleven young ladies who had come to India to make some rich men put on the matrimonial crown of thorns, disappointed their friends when they got through the surf off that port.

It had been expected that the new arrivals would have brought with them the latest fashions in hats and clothes, and indeed they had left home with a very adequate stock. Unfortunately, during the last three months of the voyage, tempers had worn rather thin—the differences of opinion among the occupants of the Round House had been so acute that when the eleven young persons landed, they could not show a single undamaged hat between them. Clothes had been torn to shreds.

All that helps to prove, in a millinery sense, the truth of the Indian proverb—'It isn't always the woman last from the well who brings the freshest water'.

It must also be remembered that woman was always a fighting animal. For a real scrap, take two women and one man!

Harry Hobbs, *John Barleycorn Bahadur*, 1943

Many [of the Fishing Fleet] who started full of hope must have bitterly regretted finding that, in the end, they had

Stayed to marry half a heart,
little more than half a liver.

Ibid.

Description of the Arrival—Amongst
Others—of the Fishing Fleet in Calcutta in 1852

What a variety! Here, for instance, are damsels young and fair sent out by knowing parents, and consigned to wide-awake relations or connections as agents for the disposal of their charms to the highest bidder. Calcutta is a famous market for such ware—European stock is always above par.

Not that these fair creatures are at all likely to be themselves passive and indifferent toys in the hands of others—far from it! We may rather regard them as jovial young huntresses trained to the chase, but instructed if possible to seize on a Judge, a Collector, or one of some other tame species, as such are the most valuable and when to be met with at liberty oft easily allured, though now and then they will offer resistance. Failing, however, to find such, they are directed, and in truth far prefer, to proceed to the haunts of the fiercer tribes, being still taught to select as the best prizes, the old and sleek that have grown fat in the rich pastures of the East, but are now casting a wishful eye to the distant coverts 'mid which they were reared and evincing a desire to retreat to these. If the fair Dianas cannot capture a General, they must be content with a Colonel, and if a Colonel does not present himself, a Major may be taken. As a last resource, Captains or even Subs—and of these there are always plenty to be caught—may be appropriated; if they should happen to be shy they must be netted and snared. But it is only the less captivating of the sex who have ever to resort to artifice; a single cast of the lasso will often noose two or three couples, and sometimes a whole herd together, while perseverance is sure to reward the most inefficient with success.

And it is to be remembered that few of the fair daughters of Europa who engage in this species of adventure, contemplate or wish for a very long sojourn in the field—in a word, they hope, when they have achieved their object, to return whence they came as the companions of pensioners, or the wealthy relics of deceased husbands, or at least often to visit their native land, sometimes in company with uxorious pards, but more frequently alone, bearing heavy drafts on London bankers.

Robert George Hobbes, *Scenes in the Cities and*
Wilds of Hindostan, vol. I, p. 6, 1852

The Griffin

When Jones arrived, a youth unwived, at Bombay's Ballard pier,
an unpretentious 'Griffin' at the start of his career,
he felt dismayed as he surveyed the loud tumultuous scene;
his ship-board friends had disappeared as if they'd never been.
A raging beast, the clamorous East, engulphed him in a trice;
no-one had warned him in advance—he would have liked advice.
A native sidled up and cried 'Is Sahib wanting bearer?
See, I have chit from gentleman I served with in Nowsheera.'
Another came and gave his name and pushed the first aside:
'Master no listen fellow here, he taking you for ride;
he Chlistian* thief, he fond of beef, he no good man for you . . .'
But, as he spoke, another bloke pushed this one out of view;
a dozen more came jostling him, each proffering a 'chit'.
Jones stood his ground and, looking round, espied his precious kit,
his trunk, his gun, his bedding roll, his hat-box and his grip
being carried off the landing wharf that lay below the ship
by coolies who were led by yet another unknown man—
no-one had warned him in advance the dangers that he ran.
He gave a yell, the rabble fell back somewhat in disorder,
as Jones gave chase to overtake the baggage room marauder:
'Hey, you, you there, how do you dare to carry off my boxes?'
he yelled and, pained, the guide explained he was the man
        from Cox's.
'But how d'you know where they're to go?' fumed Jones on
        getting nearer.
'This fellow here, he tell men where, he say he is your bearer;
he show me letter, brought from Quetta, now you see it too;
he looking after luggage first and after look for you.'
Jones took the letter, thought it better to ignore the slight—
the 'fellow' might be wrong but his priorities were right.
He might be wrong but he looked strong and fierce and rather proud,
a 'fellow' to be reckoned with—not just one of a crowd.
'Dear Jones' he read 'if you're not dead before you get this note,
Mohammad Khan, a good Pathan, will meet you from the boat.

---

*Christian

The Griffin

He may look rough, but he is tough—believe me when I say
he'll rescue you from all the touts and beggars of Bombay.
They'll turn and fly beneath his eye—he only needs to scowl—
like Pharaoh's chickens on the kill who hear the tiger growl.
It won't be rash to give him cash for tickets, tips for porters,
for early morning tea and toast, or early evening 'snorters'.
Give him his head—he'll make you bed and have your
    carriage cleaned;
he'll care for you as if you were an infant not yet weaned.
I think that's all—excuse the scrawl—be seeing you, Pip Evans'.
'Pip Evans! Why good heavens, he had fagged for him at school;
though good at sport young Jones had thought old Evans was a fool;
and like as not, Jones was a swot in Evans' estimation.
Quite fair perhaps that both these chaps should represent their nation.
He folded up the letter, feeling better now, he smiled;
Mohammad Khan, the good Pathan, no longer seemed so wild.
The fellow grinned in quick response and shook his proffered hand.
The motley crew of touts withdrew, as if by magic wand.
The man from Cox pulled up his socks and gave a nervous cough;
now all was well and he could tell the porters to move off.
So Jones progressed, his mind at rest, to catch the Frontier Mail,
and could enjoy without annoy the spell of India's rail.
No pen can tell that magic spell in which young Jones delights—
the sounds and smells, the hawkers' yells, the fascinating sights.
So, we must leave him in the care of tough Mohammad Khan;
and leave you, reader, to digest this touching little yarn.

Anon, 1928

Calling

A stranger here a week ago,
(there isn't yet a soul I know)
I went out 'calling';
in smart store suiting, as is meet,
I went at noonday and the heat
was quite appalling.

Pharaoh's chickens vultures   snorters alcoholic drink

Nowhere I saw throughout the place
one welcome glass or friendly face,
or glimpse of beauty;
and finding neither rest nor shade,
I cursed what nincompoops have made
a social duty.

Marked 'Not at Home' (was I deceived?)
each box unblushingly received
my salutation;
till, card case empty, I returned
sad for the folly that had earned
such tribulation.

A week has passed and strange but true
my knowledge as to who is who
is just as scrappy;
convention satisfied, who cares?
They have my cards and I have theirs,
so both are happy.

'Momos'

THE HEAVEN-BORN

EDITOR'S INTRODUCTION

This title was given to the Indian Civil Service by members of other
services who were jealous of their premier position and resented a
certain arrogance or conceit which they detected in some members
of the ICS. The label stuck, and some members of the service even
came to regard it with some affection. The analogy, of course, was with
the twice-born brahmin, the highest Hindu caste.

Up until 1806, the young men who were the precursors of the ICS
were selected by patronage, exercised mainly by the Directors of the
East India Company at home. They received no specific training.
Many of them came out very young to become writers, i.e. clerks. But
in the stirring times of the eighteenth century, there were plenty of
opportunities for ambitious and able young men to make their mark

in politics, administration, and leadership of various kinds. It can be asserted that the system of selection was not wholly bad, when it produced such brilliant administrators as Warren Hastings, Malcolm, Munro, Elphinstone, and Metcalfe.

In 1806 a college was started at Hertford for the specific purpose of educating the young men recruited for service in India. The college was moved to Haileybury in 1809, and lasted for 50 years.

Strictly speaking, the ICS was born in 1858 when all the territories acquired by the East India Company and all the staff they had recruited to administer these territories were taken over in the name of the Crown. In fact, of course, there was no sudden break with the past so far as the staff were concerned, only a change in the method of selection, which was now to be entirely by competitive examination, but subject to a viva voce test. Haileybury had given an education almost as good as that provided by an Oxford or Cambridge college and in some ways better. Some of the older Haileybury-trained men spoke disparagingly of the 'competition-wallahs' in the mistaken belief that eminence in learning would only fit them for excellence in desk work. Time proved them wrong. But it was true that after 1858 never again were there opportunities such as those which presented themselves to their predecessors in the turbulent eighteenth century. After 1858 the Civil Service had its hands more than full, dealing with the trials and tribulations of the rural population. Some officers tried to educate the Municipal Boards to work honestly as well as efficiently, but few had any lasting success. Indeed if the service is to be remembered it should be for its unremitting help to improve the lot of the Indian peasant.

The total number of the ICS at any one time varied between just under 1000 and 1300. There were ultimately ten provinces varying in size from 97,000 to 1,70,000 square miles. They were all larger than the United Kingdom (94,211 square miles). There were in any province at most 100 Indian civil servants on active duty in the districts at any one time and up to 50 on leave. This was very sensible in a country with such a deleterious summer climate in the north, such an enervating climate in the south. It also provided most valuable opportunities for junior officers to gain experience while officiating in senior posts. In any one province there were anything up to 30 officers on loan to the Central Government. Some were on temporary loan to particular Indian states and some had left the province for good to join the Political Service. Members of the service held nearly all the highest posts in the Government of India. In every province there were fewer districts than there are counties in England. It follows that the districts were larger than an English county. Yet one civilian with some assistance from one or two junior civilians and half a dozen experienced Indian subordinates, and working with a Police Superintendent and his Indian subordinates, upheld law and order and also dealt with the hundred and one matters which cropped up every day, in an area larger than an English county.

There was no multiplicity of officers from whom the Indian could seek redress. There was just one—the District Officer. He was their Ombudsman, but very much more. He was not superhuman but needed to be patient and shrewd and to tour widely in his district. Of course, his task was simplified by the fact that the rural population of India, by and large, were very peaceable and also by the fact that their needs were very simple. They wanted schools for their children but not libraries, seeds for their crops but not combine-harvesters, tube-wells but not drains. If they had expected services as sophisticated as those enjoyed by the population of an English county as of right, it would have been impossible to administer an Indian district with such a meagre staff.

Another thing which greatly helped the district staff was the poverty of communication. Because of this, it was difficult to refer anything which required immediate action to higher authority. So the District Officers decided almost everything themselves and the absence of interference wonderfully concentrated their powers of mind.

Up to about 1914 all District Officers and Superintendents of Police were British. By 1946, 50 per cent of all posts usually held by members of the ICS and Indian Police were held by Indian members of the two services.

*

With the Collector

Was it not the Bishop of Bombay who said that man was an automaton plus the mirror of consciousness? The Government of every Indian province is an automaton plus the mirror of consciousness. The

Secretariat is consciousness and the collectors form the automaton. The Collector works, and the Secretariat observes and registers.

To the people of India the Collector is the Imperial Government. He watches over their welfare in the many facets which reflect our civilization. He establishes schools and dispensaries, gaols and courts of justice. He levies the rents of their fields, he fixes the tariff, and he nominates to every appointment, from that of road-sweeper or constable, to the great blood-sucking officers round the Court and Treasury. As for Boards of Revenue and Lieutenant-Governors who occasionally come sweeping across the country, with their locust hosts of servants and petty officials, they are but an occasional nightmare; while the Governor-General is a mere shadow in the background of thought, half blended with 'John Company Bahadur' and other myths of the dawn.

The Collector lives in a long rambling bungalow furnished with folding chairs and tables, and in every way marked by the provisional arrangements of camp life. He seems to have just arrived from out of the firmament of green fields and mango groves that encircles the little station where he lives; or he seems just about to pass away into it again. The shooting-*howdahs* are lying in the verandah, the elephant of a neighbouring landowner is swinging his hind foot to and fro under a tree, or switching up straw and leaves on to his back, a dozen camels are lying down in a circle making bubbling noises, and tents are pitched here and there to dry, like so many white wings on which the whole establishment is about to rise and fly away—fly away into the 'District', which is the correct expression for the vast expanse of level plain melting into blue sky on the wide horizon-circle around.

The Collector is a bustling man. He is always in a hurry. His multitudinous duties succeed one another so fast that one is never ended before the next begins. A mysterious thing called 'the joint' comes gleaning after him, I believe, and completes the inchoate work. The verandah is full of fat black men in clean linen, waiting for interviews. They are bankers, shopkeepers and landholders, who have only come to 'pay their respects' with ever so little a petition as a corollary. The chuprassi-vultures hover about them. Each of these

---

Howdahs large chair carried on the back of an elephant for ceremonial use, big game shooting, transport

The Collector's Residence

obscene fowls has received a gratification from each of the clean fat men; else the clean fat men would not be in the verandah. This import tax is a wholesome restraint upon the excessive visiting tendencies of wealthy men of colour. Brass dishes filled with pistachio nuts and candied sugar are ostentatiously displayed here and there; they are the oblations of the would be visitors. The English call these offerings 'dollies'; the natives *dali*. They represent in the profuse East the visiting cards of the meagre West.

Although from our lofty point of observation, among the pine trees, the Collector seems to be of the smallest social calibre, a mere carronade, not to be distinguished by any proper name; in his own district he is a Woolwich Infant; and a little community of microscopicals—doctors, engineers, inspectors of schools, and assistant magistrates, look up to him as a magnate.

They tell little stories of his weaknesses and eccentricities, and his wife is considered a person entitled 'to give herself airs' (within the district) if she feels so disposed; while to their high dinners is allowed the use of champagne and 'Europe' talk on aesthetic subjects. The Collector is not, however, permitted to wear a chimney-pot hat and gloves on Sunday (unless he has been in the Provincial Secretariat as a boy); a Terai hat is sufficient for a Collector.

A Collector is usually a sportsman; when he is a poet, a correspondent, or a neologist it is thought rather a pity; and he is spoken of in undertones. Neology is considered especially reprehensible. The junior member of the Board of Revenue, or even the Commissioner of a division (if he be *pucca*) may question the literal inspiration of Genesis; but it is not good form for a Collector to tamper with his Bible. A Collector should have no leisure for opinions of any sort.

I have said that a Collector is usually a sportsman. In this capacity he is frequently made use of by the Viceroy and long-shore Governors, as he is an adept at showing sport to globe-trotters. The villagers who live on the borders of the jungle will generally turn out and beat for the Collector, and the petty chief who owns the jungle, always keeps a tiger or two for District Officers. A Political Agent's tiger is known to be a domestic animal suitable for delicate noble Lords

---

Terai hat   wide-brimmed felt hat, as worn by hill regiments of the Indian Army   *Pucca*   solid, genuine, hundred per cent, in this case, not just an acting commissioner

travelling for health; but a Collector's tiger is often believed to be almost a wild beast, although usually reared upon buffalo calves and accustomed to be driven. The Collector, who is always the most unselfish and hospitable of men, only kills the fatted tiger for persons of distinction with letters of introduction. Any common jungle tiger, even a maneater, is good enough for himself and his friends.

The Collector never ventures to approach Simla, when on leave. All Simla would stare and raise their eyebrows if they heard that a Collector was on the hill. They would ask what sort of a thing a Collector was. The Press Commissioner would be sent to interview it. The children at Peterhoff would send for it to play with. So the clod-hopping Collector goes to Nainital or Darjiling, where he is known either as Ellenborough Higgins, or Higgins of Gharibpur, in territorial fashion. Here he is understood. Here he can babble of his *bandobast*, his *balbacha* and his *bawarchikhana*; and here he can speak in familiar accents of his neighbours, Dalhousie Smith and Cornwallis Jones. All day long he strides up and down the club verandah with his old Haileybury chum, Teignmouth Tomkins; and they compare experiences in the hunting-field and office and denounce in unmeasured terms of oriental vituperation the new sort of civilian who moves about with the penal code under his arm and measures his authority by statute, clause, and section.

In England the Collector is to be found riding at anchor in the Bandicoot Club. He makes two or three hurried cruises to his native village where he finds himself half forgotten. This sours him. The climate seems worse than of old, the means of locomotion at his disposal are inconvenient and expensive; he yearns for the sunshine and elephants of Gharibpur, and returns an older and a quieter man. The afternoon of life is throwing longer shadows, the Acheron of promotion is gaping before him; he falls into a commissionership; still deeper into an officiating seat on the Board of Revenue. *Facilis est descensus*, etc. Nothing will save him now; transmigration has set in; the gates of Simla fly open; it is all over. Let us pray that his halo may fit him.

G.R. Alberigh-Mackay ('Sir Alibaba K.C.B.'),
Twenty-One Days in India, 4 October 1879

Bandobast household management, arrangement for any function   *balbacha* children   *bawarchikhana* (*bobberjee-khana*) kitchen

On Assistant Magistrates

No. 11: All Assistant Magistrates on their first arrival in this country, stuffed like Christmas turkeys with abstracts and notes, the pemmican of school-boy learnings, are more or less a weariness and a bore; but the youth who comes out from the admiring circle of sisters and aunts with the airs of a man of the world and the blight of a premature ennui is peculiarly insufferable

Idem, *The Orphan's Good Resolutions*, 1880

Vale atque Ave

The Deputy Commissioner was leaving for Bombay,
and grief and *pan-supari* were the order of the day.

The members of the District Board flocked in from far and near
to garland him with marigolds and shed the silent tear.

DC's in plenty there had been, and many more would be,
but never such a Deputy Commissioner as he.

The people prayed that Providence some day might send him back;
the thought of his departure made the very sun look black.

The welfare of the District was entirely due to him;
his loss was on a par with amputation of a limb.

He was the rudder of their ship, their Captain and their Cox—
when he was gone the District would be dashed upon the rocks.

Their beacon he, their Oriflamme, their harbinger of Light,
the Star that guided them among the pitfalls of the night.

They brought him spices, scent and pan and flowers his form to deck,
and photographed him standing with the wreaths around his neck.

The train steamed out and someone cried 'three cheers and hip, hurrah'
and they felt that the proceedings had gone off with great eclat.

So he was gone—their Cherisher and Friend and Wrong-redressor—
and a similar *tamasha* was arranged for his successor.

'Bhusa'

Pan-supari combination of betel leaf and areca nut, offered to guests after a meal—often a polite intimation of the termination of their visit   *tamasha* 'show' of almost any kind and anywhere

The Aged Joint

While I was searching Hindustan
  for something to appoint,
I met an aged, aged man
  officiating 'Joint';
right lean he was, and meanly clad,
  his beard was long and grey;
I asked him why he looked so sad,
  and how he earned his pay.

'I sit in court from morn to eve,
  recording every word
of evidence I don't believe,
  on what has not occurred;
and if I give my feelings vent,
  or fail to dot my 'i's
some gentleman in Parliament
  is sure to criticize.'

. . . .

'I sometimes sit on Local Boards,
  and, while the members doze,
I teach them rules for grants to schools,
  designed by Mr H_Se;
I find them land to build a pound,
  or funds for training *dai*s,
and thus the Board becomes renowned
  for local enterprise.'

. . . .

'I number crops and toddy shops,
  I give out doles for wells,
I search the banks of village tanks
  and drains for horrid smells;
I soothe the frequent village brawl,
  and help to poison rats—

Joint Joint Magistrate   *dai* midwife

Such tasks I should not mind at all,
but after twenty years they pall
    on stiff-necked bureaucrats.'

. . . .

And now, when it is asked of me,
    and I am unprepared
to say that warders always see
    the convicts' blankets aired;
or when some ill-informed MP's
    attempt to make a point
by asking whether mughs, or thugs,
    are ever licensed to sell drugs,
or whether sweepers in Orai
    are all in debt, and if so why—
I think of that old man, whose knees
    are bent at forty-five degrees,
who always does his best to please,
    who speaks in a perpetual wheeze,
whose scanty pittance of rupees
    is insufficient to appease
the hungry cries for bread and cheese
    of infants clustering like bees
around his humble board, while he's
    officiating Joint.

                    C.H.B. Kendall, ICS,
     The Aged Joint and Other Verses, 1922

On the Indian Civil Service

The capacity of a civilian's mental power should be similar to that of
the elephant's trunk, which can pick up a pin and pull down a mighty
forest tree.

                                 C.T. Buckland

Facilis Descensus Averno

I met a chap one day in Hell
and asked him how he liked it: 'Well,'
he answered candidly, 'you see,

Mughs Burmese cooks, much sought after

there's nothing strange in this to me.
I lived among them all—the heat
the stink, the racket, the deceit,
corruption, avarice and fear;
and nearly all my friends are here.
It suits me well, I must confess;
for I was in the ICS.'

D.H.A. Alexander, ICS, 1940s

II Ballade of the Joint in Harness

The burden of Inspections. Evermore
    thou shalt ride forth, sleepy and shivering,
to question and examine and explore,
    and poke a prying nose in everything.
    To sit with village elders in a ring
or *janch-partal*ling through the fields perspire,
    and so at noon-tide to thy fast-breaking:
This is the end of every man's desire.

The burden of *cutcherry*. Now give o'er
    thy hasty pipe, let be thy cud-chewing:
go, try a dozen cases or a score
    'twixt thy down-sitting and thine up-rising.
    Then shalt thou in five minutes strive to wring
such judgments out that, as they go up higher,
    no leisured court may have at thee a fling:
This is the end of every man's desire.

The burden of much *peshi*. To thy floor
    basta on basta, shall the *ahlmad*s bring
things new and things most ancient from their store,
    and daze thin ears with drowsy murmuring.
    Then many a file to languid life shall cling,

---

*Janch-partal*ling checking a percentage of all village registers kept by the revenue accountant, covering ownership tenancy rights, crops   *cutcherry* building at the district H.Q. which housed the District Officers, the Treasury and Magistrates' Courts   *peshi* paperwork   *basta* cloth wrapped round a file (or files) to keep out the dust   *ahlmad*s Indian clerk

whose tardy fate draws nigh and ever nigher,
    till *dakhil daftar ho* its requiem ring:
This is the end of every man's desire.

              Envoi
    Prince, there is pleasure in the life I sing
when chairs are drawn to the post-prandial fire,
    till sleep effaces all with quiet wings:
This is the end of every man's desire.

                    A.G. Shirreff, ICS, 1918

WOMEN

EDITOR'S INTRODUCTION

This is a very wide-ranging and complex subject and though I have collected quite a number of items, I am not satisfied that they cover the whole subject. The women written about were all British, except for one Eurasian (who considered herself British). But the items I have collected for this anthology do not include teachers, missionaries and the wives of British Other Ranks (BORs). This is a pity, as missionaries and teachers in particular probably got closer to the real India than the wives of gazetted officers who, with some exceptions, only came into contact with their servants, a few Indian shopkeepers, some Indian gazetted officers and their wives, and, very occasionally and in a very superficial manner, with leading Indian citizens and sometimes with Indian nobility. The reason why I have had to omit the teachers, etc. is that I was unable to find any writing by or about them which I considered suitable. As for the BORs' wives, they have in Kipling their inimitable champion. All British women, whatever their origin and upbringing, had to adapt themselves to a new way of life and some were more successful at this than others.

    The British women in India dealt with in this chapter could be divided into three groups:

Dakhil daftar ho   deposit in the record room of the government office

1. A few who loathed everything about India and only wanted to leave for good as soon as possible. They could go home while their husbands stayed on in India. This almost inevitably led to the break-up of their marriages. Or the wife could sacrifice herself out of love and loyalty for her husband and stay with him. This was also likely to lead to the break-up of their marriage, because of the great strain imposed on both of them. Or, her husband might resign from service in India, take his wife home and try to find another job. The key to the survival of this marriage would probably be his getting at least as good a job in the United Kingdom as he had held in India. In my experience this last solution was hardly ever resorted to.

2. A much larger number who loved the life in India. These were found most frequently among those who took full advantage of the sports that India had to offer—tennis, riding, hunting, shooting, fishing, camping, trekking, sailing, ski-ing, and perhaps for all one should add dancing and entertaining. Most of these women were young. As they grew older, most of them would join the third group described below.

3. Those who in the North enjoyed the cold weather but dreaded the hot weather and rains in the plains, and those in the South who disliked the hot moist weather lasting through most of the year, but who steeled themselves in either case to endure with the help of one or two short visits to the hills. These were the hardier souls. But in the North it was not at all unusual for wives with the approval of their husbands to stay up in the hills for the full five months from May to September. If there were small children it was essential to keep them in the hills for the full five months. Some of the women in this group found they had reserves within themselves which made it easier to put up with the climate, the stupidity of servants, and the confidences of younger women, which they did not want to hear. Such women were indeed lucky and so were their husbands.

A young man was well advised to try and make sure that his fiancée would not be in the first group, even if this meant arranging that she should come out for a year to stay with her own or his friends, before they got married. This ideal advice was seldom followed and was not necessary in the case of girls who had already spent some time in India.

Kipling had little use for the average well-bred genteel woman in India. They were dull and provided him with no 'copy'. Instead he wrote with uncanny insight about the wives of BORs and amusingly of clever, witty and malicious women, who haunted the hill stations and who took pleasure in being sometimes kind but more often odious. As a result his readers at home got the impression that most *memsahibs* were as portrayed by this champion of Empire, which was of course quite untrue. For every Mrs Hauksbee there were a hundred kind and generous women with varying gifts and abilities. If the British lost their close association with Indians as a result of the arrival of British women in India, it was a price well worth paying for the companionship and support they gave to their own menfolk.

The greatest enemy of women in India was boredom. It drove them to drink, to flirt, to hypochondria, to quarrel, but seldom, it seems, to 'scribble' like their menfolk. There were several reasons for this. They were not as well educated, at least in the skill associated with Latin and Greek and English verse, and there was still a very strong Victorian tradition that women should content themselves with writing letters, diaries and perhaps some romantic poetry. But the strongest reason lay in the very nature of women and its difference from that of men. It will not have escaped your notice that a very large proportion of the items in this anthology are devoted to ridicule and satire. Most women prefer to wage war with words and looks rather than with the pen. The spoken snub, the unspoken look of scorn or contempt, the barbed repartee—to perfect these weapons was at least as difficult as writing a satire, and much more satisfying. So we have men writing satirically about women, without any reply. But, given a *tête-à-tête* encounter, I'd back the woman to win every time.

*

Chit-Chat

One of my correspondents some time ago very ingeniously discussed this disposition to letters, though perhaps with too much severity. Whether in the petty but constant and universal manufacture of 'chits' which prevails here, it deserves his appellation of 'cacoethes scribendi'

Memsahib English woman, its shortened version being *mem*.

must be left to the judgment of the learned. But be it worthy of praise or blame, the fact is certainly notorious that in no part of the world is to be seen in more practice and perfection the art of 'scribbling', and, as another of my auxiliary friends remarked, that manners and language keep pace with each other—it is observable that in this favourite instance an appropriate phrase has been very expressively applied. The small scale on which this literary intercourse is carried on, naturally suggests a pretty little infantine idea. The term 'chit', therefore, which is well known to mean nothing more or less than a little child, is very properly adopted to signify this baby correspondence.

The superiority of talent of the gentler sex in all the pleasurable and elegant intercourses of life is universally acknowledged; and not in any instance more than in the facility of taste with which they communicate their ideas both in conversation and correspondence. This double power is happily expressed in the compound epithet descriptive of the two arts in which they excel—'chit-chat'; evidently importing the ready facility both of pen and tongue.'

Hugh Boyd in *The Indian Observer*,
no. XXVIII, 18 March 1794

Miss Emily Eden to Mrs Lister

Government House
Calcutta
25 January 1837

My dearest Theresa, I will take your plan of sitting down forthwith and answering your letter (of August 18th, received January 23rd) on the spot, before the pleasure of reading it wears off. It means I am going to answer your letter directly, and I am so obliged to you for asking me questions—just what I like. Intellect and memory both are impaired, and imagination utterly baked hard, but I can answer questions when they are not very difficult, and if they are put to me slowly and distinctly; and besides, I am shy of writing and boring people with Indian topics. I used to hate them so myself. But if they ask about an Indian life, as you do, and about the things I see every day, why, then, I can write quite fluently, and may heaven have mercy upon your precious soul! So here goes:

'Do you find amongst your European acquaintances any pleasing or accomplished women?' Not one—not the sixth part of one; there is not anybody I can prefer to any other body, if I think of sending to ask one to come and pay me a visit, or to go out in the carriage; and when we have had any of them for two or three days at Barrackpore, there is a *morne* feeling at the end of their visit that it will be tiresome when it comes round to their turn of coming again. I really believe the climate is to blame.

'They read no new books, they take not the slightest interest in home politics, and everything is melted down into being purely local.' There is your second question turned into an answer, which shows what a clever question it was.

Thirdly. It is a gossiping society, of the smallest macadamized gossips I believe, for we are treated with too much respect to know much about it; but they sneer at each other's dress and looks and pick out small stories against each other by means of the *ayah*s, and it is clearly a downright offence to tell one woman that another looks well. It is not often easy to commit the crime with any regard to truth, but still there are degrees of yellow, and the deep orange woman who has had many fevers does not like the pale primrose creature with the constitution of a horse who has not had more than a couple of agues.

The new arrivals we all agree are coarse and vulgar—not fresh and cheerful, as in my secret soul I think them. But that, you see, is the style of gossipry.

An Extract

The Irrepressible Mrs Parkes

We are rather oppressed just now by a lady, Mrs Parkes, who insists on belonging to the camp. . . . She has entirely succeeded in proving that the Governor-General's power is but a name. . . . The Magistrate of one station always travels on with us to the next. To each of these magistrates she has severally attached herself, every successive one declaring they will have nothing to do with her. Upon which G. observes with much complacency: 'Now we have got rid of our Mrs P'—and the next morning, there she is, on the march, her fresh victim driving her in a tilbury and her tent pitched close to his. . . .

Ayah lady's personal maid, and nanny for children

I am sure you will be pleased to know that yesterday, as we were returning from ruins, G. said: 'Here come Macintosh and Colvin on an elephant; how fat Colvin grows!' Colvin turned out to be Mrs Parkes in a man's hat and riding habit. She had met Captain Macintosh and, as far as we can make out, climbed up the elephant's tail into his howdah when least expected.

She will certainly be the death of us all.

Ibid.

Three and–an Extra

Then said Mrs Hauksbee to me—she looked a trifle faded and jaded in the lamplight—'take my word for it, the silliest woman can manage a clever man; but it needs a very clever woman to manage a fool.'

Kipling, *Plain Tales from the Hills*, 1888

My Rival

I go to a concert, party, ball—
what profit is in these?
I sit alone against the wall
and strive to look at ease.
The incense that is mine by right
they burn before her shrine;
and that's because I'm seventeen
and she is forty-nine.

I cannot check my girlish blush,
my colour comes and goes;
I redden to my fingertips,
and sometimes to my nose.
But she is white where white should be,
and red where red should shine.
The blush that flies at seventeen
is fixed at forty-nine.

I wish I had her constant cheek;
I wish that I could sing
all sorts of funny little songs
not quite the proper thing.

I'm very gauche and very shy,
her jokes aren't in my line;
and, worst of all, I'm seventeen,
while she is forty-nine.

The young men come, the young men go,
each pink and white and neat,
she's older than their mothers, but
they grovel at her feet.
They walk beside her rickshaw wheel—
none ever walk by mine;
and that's because I'm seventeen
and she is forty-nine.

She rides with half a dozen men,
(she calls them 'boys' and 'mashers')
I trot along the Mall alone;
my prettiest frocks and sashes
don't help to fill my programme card,
and vainly I repine
from ten to two a.m. Ah me!
Would I were forty-nine.

She calls me 'darling', 'pet' and 'dear'
and 'sweet retiring maid'!
I'm always at the back I know,
she puts me in the shade.

She introduces me to men—
cast lovers I opine,
for sixty takes to seventeen,
nineteen to forty-nine.

But even she must older grow
and end her dancing days.
She can't go on for ever so
at concerts, balls and plays.
One ray of priceless hope I see
before my footsteps shine;
just think, that she'll be eighty-one
when I am forty-nine.

Kipling, *Departmental Ditties*, 1886

Triolet

It turned to a kiss,
      I intended a quarrel.
Flirtatious young miss!
      (Yet it turned to a kiss.)
She pouted—and this
      robs my verse of a moral.
It turned to a kiss;
      I intended a quarrel!

                    Sir J.A. Thorne, ICS

The Plaint of Thestilis

Cicala courts cicala; through the leaves
chameleons whisper to their vis-à-vis
and every morn I hear beneath the eaves
a hymn of wedlock in a myna key.
Ah! Mother of Love! Bring back my man to me!

Once 'twas a subaltern I hoped to wed,
Cupid in khaki, Love's facsimile,
who carried 'Thestilis' tattooed in red
upon his arm for all the world to see.
Ah! Mother of Love! Bring back my man to me!

And then a victim of the brown and blue,
I tracked the footsteps of a D.S.P.
who held me rapt with tales of derring-do,
who gave me silk and silver filigree.
Ah! Mother of Love! Bring back my man to me!

Sad is my lot, yet still the memory smiles
to think of him—that wan and pale D.C.,
who crept bewildered through a world of files,
yet once found time to take me on his knee.
Ah! Mother of Love! Bring back my man to me!

For five long years I've fought through thick and thin,
and next July I shall be twenty-three,

so he be padre or a son of sin,
a man of parts or partial pedigree,
Ah! Mother of Love! Bring back some man to me!

> J.M. Symns, Indian Educational Service,
> *The Mark of the East*, 1913

The Mark of the East

When Gertrude sails for India,
she bids her kith and kin
inspect the bales of tropic veils,
the helmets made of pith:
the net to spread above her bed
is viewed with anxious mien,
and eyes dilate to see the crate
of camphor and quinine.

When Gertrude sails for India,
her mother's feeling queer,
the Rector blows an anxious nose
and wipes away a tear:
shall Ruth or Grace usurp the place
'tis Gertrude's right to hold
in Little-Budleigh-in-the-Mud-
cum-Worple-on-the-Wold?

When Gertrude sails for India,
the local 'Dorcas' sighs
for one whose zest last autumn dress'd
a score of pagan thighs.
In stricken tones a curate drones
the lessons for the day,
nor dares to view the Rector's pew
for fear of giving way.

When Gertrude comes from India,
she's Indian to the core,
her gown and hair, her manners bear
the stamp of Barrackpore.
She sits and prates of maiden plates,

of revels at the 'Gym',
of leading parts and doubled hearts,
the regiment and him.

When Gertrude comes from India,
she's found an Eastern twang,
and bores her friends with odds and ends
of Anglo-Indian slang.
The roof-tree shakes, the housemaid quakes
before that horrid flow
of *idhar ao* and *jaldi jao*,
and *asti bat karo.*

When Gertrude comes from India,
the Rector's habits pall,
the startled guest is gently press'd
to cocktails in the hall.
Her parents pale before the gale
which swamps the old routine,
and, save in Lent, must needs consent
to dine at 8.15.

When Gertrude comes from India,
the schemes I'd lately planned,
they fade and die, and that is why
I loathe that selfish land,
which drains the West of all its best
to keep an atlas red;
which dared to claim my only flame
and sent me this instead.

Ibid.

Mrs Plantagenet Paley

Mrs Plantagenet Paley
was thought by some young men divine,
so she was to sit at the table
where the Viceroy was going to dine,
for 'twas said that his Lordship had flirted on board ship,
and liked to see pretty eyes shine.

Idhar ao come here   *juldi jao* go quickly   *asti bat karo* speak slowly

Mrs Plantagenet Paley
was clever besides, and she thought:
'If the Viceroy could only do something
for Jack'—so she felt that she ought
to be lively and pleasing while others were freezing—
She was, and she seemed to have caught

the great man's attention as often
he smiled on her sweetly and said
pleasant nothings, which some distant ladies
felt sure must be turning her head;
for Viceregal smiling begets much reviling
in quarters where it is not shed.

After dinner there came a reception,
H.E. stood close to the door
and bowed to the world as it passed him,
paying compliments, yes, by the score
to ladies whose beauty demanded such duty,
which to him was by no means a bore.

An hour had passed when his Lordship
thought he saw a new beauty and straight
advanced and said 'avec effusion'—
'Ah, cruel to make us all wait
such a long time despairing of seeing you wearing
that beautiful dress—Why so late?

'As we droop you arrive in your freshness,
too bad our hopes thus to deride;
we have longed for your presence'—the Lady
addressed, her blue eyes opened wide—
for her name it was Paley, the same who had gaily
at dinner conversed by his side.

Moral

So ladies if e'er you imagine
you have charmed an old diplomat, stay
to reflect on the truth of the moral

illustrated once more by this lay—
the moral which teaches that Viceregal speeches
are sometimes forgotten next day.

'Lunkah', *Whiffs Anglo-Indian*, 1891

The Ladies of Mahabaleshwar

The ladies of Mahabaleshwar
they are so fresh and gay,
they have the Kruschen feeling
most wonderfully all day.
But Poona! Oh in Poona
they languish there and swoon,
and get heat apoplexy
upon their beds at noon.

The ladies of Mahabaleshwar
have strawberries for tea,
and as for cream and sugar,
they add them lavishly.
But Poona! Oh in Poona
their hearts are like to break,
for while the butter's melting,
the flies eat up the cake.

The ladies of Mahabaleshwar
are never known to yawn,
although they dance till midnight,
and ride out with the dawn.
But Poona! Oh in Poona
they creep to bed at ten,
and if they wake for breakfast,
well, no one asks them when.

The ladies of Mahabaleshwar
like Grecian nymphs are seen,
tiptoe on airy uplands
among the woods so green.
But Poona! Oh in Poona

it's torture to their feet
if they should chance to venture
across the burning street.

The ladies of Mahabaleshwar
in wraps and furs delight,
and often get pneumonia
'neath blankets two at night.
But Poona! Oh in Poona
the gauziest wisps appal,
and ladies sleep (they tell us)
with nothing on at all.

The ladies of Mahabaleshwar
in such sweet charms abound
that doctors say their livers
are marvellously sound.
But Poona! Oh in Poona'
they scold and nag all day,
and contradict their husbands
until they fade away.

'Momos'

The light that lies in Women's eyes
and lies and lies and lies.
. . . .

Two young girls staying at Old Delhi's most famous hotel, sent this telegram to their parents:

'All money spent; can no longer stay maidens.'

Harry Hobbs,
John Barleycorn Bahadur, 1943

The Lady Who Has Never Been Home

Our Colonel was a great dandy ever so many years ago, when he was an Ensign (and Lieutenant) in the Foot Guards for about a twelvemonth, during which he acquired that elegant languor of manner which made him so distinguished in Indian Society, and that fine crop of debts which prevent his transferring his once charming presence elsewhere. But Colonel Fugleman's debts are not the only ties that retain him in this land of exile; the rosy fetters of Cupid (or rather the coffee-coloured bands of Hymen) are also for something in the matter.

The fact is that Fugleman, though a fine soldier, was a fop; and, during the Mutiny, was much flattered by the attention which he received from the daughter of a planter—Miss Tommkyns of Tomm-kynsabad—whose life he saved by a brilliant night march. Miss Tommkyns, at the time of the romantic adventure referred to, was a very pretty girl—dark but comely as the tents of Kedar. She had been educated at Calcutta, and had a pronunciation of her mother (?) tongue which was elaborate rather than idiomatic. She described, in those artless accents, to a friend her first meeting with Fugleman as follows:

In thee morning, ass sun as it was light, there we saw the Kunnal and, oh my! didn't he look splendeed? Of course, you know, we were riding in the Buggee, and he was on hees horse, in the center of the rod and when we came by, of course, you know, hees horse commenced to hit a keek, and, oh my! wee got so frightened.

The upshot of it was that Colonel Fugleman lost his seat and his horse fell upon him. His left leg was fractured in two places, and he had to be nursed at Tommkynsabad; the end was, 'of course' (to use Madeline's favourite phrase) that they became man and wife. I dare say they were happy—I am sure they were very fruitful; and beauty such as Mrs F.'s—always of a fugitive type—has been further shaken by family cares. I doubt whether any of Fugleman's former friends would recognize the ex-Guardsman in the bronzed and grizzled sloven—except on parade, where he is still smart enough—who sits with a cheroot between his lips, and a glass of gin and water before him, surrounded by tawny offspring of all ages; while his draggled, down-at-heel partner smokes her hubble-bubble or chews pan.

One good point Mrs F. certainly has—nay many; but the one of which I am now thinking is that, while regarding Mrs de Murrer as the admitted leader of fashionable life in our station, she does not share the notions of 'the Judge's ladee' on the subject of social

---

Hubble-bubble   hookah

Fugleman's Home

distinctions. That the wife of a Bengal Magnate of twenty-five years' standing should cringe and suppress herself before young Blueskin because of his presumptive claim to the Jolinose (Irish) peerage, is to her inconceivable. 'Why, that young boy' says the Colonel's wife, 'was only few days-like ready to be our assistant at thee Factoree. What is a lord? Will he ever be Viceroy? I don't think so, not even of Madras or Bombay. I will give him such a set-down just now.'

There is a great deal of difference between Mrs Fugleman when making these unaffected confidences to her old friends—generally the youngest officers of the regiment, where she is a great favourite—and the self-same lady when giving a dinner party, or what in her less-guarded moments she calls a *saddlee-peeroo khana*. She has a consummate knowledge of the poultry-yard, and manages the mutton-club for the mess; and were it not for a foolish fancy for hermetically sealed provisions and a hopeless indifference to the quality of her wines, she would not give at all a bad feed. This craving for the secrets of the Trismegist is by no means confined to little brown Madeline, as many diners know to their cost; but she is, I have reason to believe, the original of the famous story I am now about to repeat.

Blueskin, before joining us, had been in the 8th Dragoon Guards, or Ducal Skewballs—his father, the old planter, having, with equal wisdom and good nature ceased to combat the youth's infatutation that cavalry was most suited to his appearance and pretensions. He soon got tired of the Skewballs—which was just then a perfect 'trades union' or collection of nouveaux riches—and exchanged (bad luck to him) into ours. When he first joined he called, 'of course', on the Fuglemans; and Madeline, who had known him as a boy, was very gracious. 'Well, Joe, how did you get on at home? I hear you were in thee Guards. Oh my! and did you often dine at Windsor Castell when you were on dutee?' Joe replied that he had dined very frequently at the Royal Table (a most ungentlemanly falsehood—but let that pass). 'And what did you get, Joee?' asked the impatient lady: 'Everything hermeticalee sealed of course?' . . .

The first Roughs have been here a good while now and the amiable Commandantess is pretty well-known, and has her own way with the tradespeople in cantonments, European as well as native. But they

---

Saddle-peeroo khana a meal consisting of saddle of mutton served with roast peafowl—an expression not likely to be used by 'pucca ladies'.

have found out the good lady's weakness in the matter of wine, and I assure you it is no joke to go to one of her dinners in the hot weather when the throat of man is naturally dry, and his head and other organs unnaturally sensitive.

De Murrer, the Judge who goes in for bland urbanity, complimented her—the humbug—on her champagne at one of the first of her dinners that he had the happiness to attend.

'Oh my! Judge' simpered the matron: 'well, of course you know it ought to be good, for we got it in our own bazaar.'

A general laugh went round the table as the ladies rose after this sally. 'Solvuntur risu tabulae', whispered the man of law to me, and never returned either to the subject or to the house. . . .

Madeline looks forward to going to England (home as she persists in calling it, though really and truly she could go to her home in less than twenty-four hours). Some one asking her how she expected to get into London society, she answered: 'Oh! it will be very easee; we will give a dinner, don't you know, to the whole station.'

H.G. Keene, ICS, *Sketches in Indian Ink*, 2nd edition, 1891

Mem-sahib

Any morning you may meet her
 where the sunlight gilds the strand
and the curlews rise to greet her
 as she gallops o'er the sand,
riding swift, as though a wager's
 in the forefront of her mind,
with a brace of breathless majors
 close behind.

Watch her dole the daily rations,
 watch her scan the butler's book,
watch her foil the machinations
 of a swart and bearded cook;
prouder than a queen, sublimer
 than a goddess, see her stand
with a Hindostani Primer
 in her hand!

Mem-Sahib

When the swift and welcome gloaming
    shrouds the palm-trees and the huts,
and the bullocks, slowly homing,
    loom like ghosts across the ruts;
when the plantain (or banana)
    rocks to rest the drowsy midge,
she'll be up at the gymkhana
    playing bridge.

And it seems a little funny
    that not one among us all
ever danced the 'Hugging Bunny',
    or the glad 'Crustacean Crawl'
till she came out East and taught us
    every trick of pose and gait,
occidentalized and brought us
    up to date.

And our bungalows were gloomy,
    there were bats behind the doors,
and the rooms were far too roomy
    with their bare and shameless floors,
till she burst upon our quiet
    with her china and her prints,
with the reminiscent riot
    of her chintz.

Would you learn the gladness of her,
    catch the charm before it pass?
Ask the butterflies that hover
    emerald o'er the sun-burned grass;
ask the paddy-birds that settle
    on the crimson-flowering boughs,
or the frangipani petal
    in her house.

And I would not have you grudge her
    any pleasure she may wrest
from the wilderness, or judge her
    by the standards of the West;

she's a 'bold designing creature'
    to the folk who know her least,
but to us—the saving feature
    of the East.

J.M. Symns, Indian Educational Service,
    The Mark of the East, 1913

Excerpts from Gone with the Raj

Towards the end of September, 1923, at the onset of the cold weather, the telegraph man appeared again—a case of blackwater fever at Ledo toward the Naga Hills. I just managed to catch the train and reached my destination about five o'clock that evening. At that time Ledo's special importance was as the centre of a small coalfield. Above ground it was surrounded by tea gardens. Later, during World War II, it was the start of the Stilwell Road to Burma and China.

The tea garden doctor who had sent for me was on the station platform. He pointed to a bungalow on the top of one of the neighbouring hills.

'That's the place, Sister. Mrs Sprott's the patient, wife of a miner. She is very ill indeed.' He indicated an overhead line of travelling buckets.

'You'll go up in one of those coal buckets. It's the only way to get there.' He lowered his voice.

'The coffin's been made. You can sit on it in the bucket. It'll be needed to-night.'

'Anything to oblige,' I said tersely and determined I'd do my best to see it wasn't. The doctor said he would be up in the morning.

After my ride in the bucket I deposited the coffin on the verandah, the husband showing no surprise. When I entered the patient's room I thought I had never seen anyone look quite so ill. Mrs Sprott lay with her eyes closed, her frail hands folded outside the bedclothes. Her long dark braids of hair accentuated the pallor of her face. The husband and the Indian dispenser hovered anxiously. A few orders and directions had been left for me by the doctor and I quickly got to work. I forget the details but I know I worked unceasingly on Mrs Sprott for forty-eight hours. The doctor called as promised. When he saw Mrs Sprott was still alive he could not believe his eyes. On the evening of the second day I suddenly saw the patient had her eyes wide open and fixed on me. I went over to the bedside to hear her whisper:

'There's one thing you ought to know. I do like a hegg with my tea.'

Soon after my return to Dibrugarh (in 1924) I was called to a case at Nazira. The patient was an elderly planter suffering from delirium tremens. His wife had died two years before and this bereavement had shattered him. He took to lonely drinking to drown his sorrow. In the various sheds and godowns on the tea gardens under his management, he secreted whisky for use during his rounds. Everyone was sorry for him yet by now there was little one could do except humour him and make him as comfortable as possible. He became mentally unhinged and partially paralysed from neuritis. He suffered from hallucinations. By the time I reached Nazira, all alcoholic drink had been denied him and he seemed to have lost his craving for it. When, after a week or two, no improvement in his condition took place, the French doctor in charge of the case decided to take our patient to Calcutta's Presidency General Hospital for treatment. It was most uncommon to find a French doctor on a tea garden—the individual tea garden managements made their own selection.

I was to accompany the party. It required all our ingenuity to persuade the patient to agree to the move. He had for some days had the delusion that a war was in progress and that the Viceroy was hiding on our roof with stores of arms and ammunition. I exploited this delusion, telling him that on instructions from the illustrious gentleman on our roof, we were to make a get-away. My ruse worked splendidly; he and I exploded with laughter at times and I have never met such an amusing and cheerful DT patient, though he had his morose moments too. Our patient began to co-operate although we had some difficulty in persuading him onto a stretcher and after a journey by lorry, onto the train.

What a journey it was, bottled up in a four-berth compartment for nearly forty-eight hours with a lunatic and an excitable Frenchman. To make matters worse, the forefinger of my right hand was jammed in the carriage door as the train left Nazira station. The pain was excruciating but fortunately we obtained ice at the next station and this helped to ease my pain. Then the doctor somehow managed to let a 'poke', the local name for an insect, creep into his ear. He became almost as mad as his patient. We had no oil amongst our medical equipment and so I put some Eno's Fruit Salts down his ear and topped up with water. This floated—or blew?—the foreign body out of his ear.

It was at one of the halts at a station during our journey that I was called to another carriage where a man travelling to Calcutta with his wife and baby, had been hiccoughing for forty-eight hours non-stop. He was extremely exhausted. I did not know how to stop this distressing hiccoughing, so, dashing back to our carriage, I sought advice from the already over-excited doctor. When he partly understood the situation down the line and learnt a baby was in that carriage, he at once said, 'Gripe water! Baby's gripe water.' I ran back to the stricken family, seized the bottle of gripe water (this useful aid always accompanied babies wherever they went) and poured half the bottle down the patient's throat. The next stop happened to be in two hours' time, so I had leisure to sit and watch the effects of my treatment. It worked well and I left a happy family at the next stop and rejoined my crazy couple down the platform.

Most of the time the patient lay on his bunk trying to catch non-existent red spiders. Heavy rain had made the usual route to Calcutta impassable and this necessitated a detour by rail through the Jaintia Hills to the Brahmaputra at Chandpur where we transferred to a steamer bound for Goalundo, some eight hours distant. Here we again entrained, this time for Calcutta. As I tucked the patient up in bed at the Presidency General Hospital two days after leaving Nazira, he said demurely: 'Thank you so much for bringing me back to school.' He died the following day.

Emma Wilson, Kaiser-i-Hind Gold Medal,
Chief Lady Superintendent,
Lady Minto's Indian Nursing Association, 1921–48

GENERALS (AND OTHER RANKS) AT RISK

EDITOR'S INTRODUCTION

Generals figure rather prominently in this section. This was partly because there were, or seemed to be, so many Generals in India. In 1935 the strength of the peace-time Army in India was approximately

150,000. Including all Auxiliary and Indian States Forces the total would have been something over 200,000. There were 55 Generals. The Army in India was organized in four Commands—northern, western, eastern and southern—covering practically the whole sub-continent. Units of this Army, large or small, were posted in 83 garrison towns throughout India. The idea was that each garrison could be inspected without too much trouble from its own Command Headquarters. Incidentally, this arrangement provided admirably for the Army's deterrent role in the preservation of internal security. Another factor to be kept in mind when considering the number of Generals is that the Army in India was composed of (a) units of the British Army doing a spell of duty in India and (b) the Indian Army. Each had their own lists and speed of promotion. Of the Generals mentioned above just under half were British Army service and just over half were Indian Army service. I have no doubt that all the Generals were usefully employed but there is no way of finding out if it was really necessary to post Generals to all these posts.

In other (hot-blooded) countries Generals were expected to lead or oppose revolutions. Our Generals were not allowed to do that. They had plenty of time to ride, hunt, shoot, play snooker and polo and bridge and, of course, to dance. They were not the only officers to dally with the juniormost subaltern's wife, but, when they did, they were inclined to use their rank to cut out other competitors. Skills in the arts of love were often accompanied by skills in the arts of war. I like to think that Kipling's General 'Bangs' in addition to having a generous sense of humour, was also a good soldier. Berryman's General was a petty-minded bureaucrat; I don't think he can have been a good soldier. Colonel Cantilever was not the only officer to stand in little awe of his General. There is the famous story of a General, who had appointed his own son, only recently commissioned, as his ADC—a not uncommon practice. One day when the General was due to inspect a famous Indian Cavalry Regiment, he sent his son with a verbal message to the Adjutant. The boy rendered the message as follows: 'Father says he may be a little late.' 'Oh, he does, does he,' said the Adjutant sarcastically—'and what does Mother say?'

The Shooting Regulations are a caricature of the actual regulations, which were nearly as unworkable. The Army Chiefs encouraged interest on the part of BORs in shooting but gave them no advice beyond saying that they must be accompanied by an interpreter and a list of the animals they must not shoot and a caution against shooting near temples or other religious buildings. In my experience the so-called interpreter was often a rogue and not to be trusted to look after a party of ignorant BORs. The only satisfactory way to educate BORs into the proper way to conduct a shooting expedition in India, was for an officer of their regiment to take them out as his guests or apprentices. Without realistic safeguards there was not a large but a steady toll of 'bloody murders' with the inevitable demand for compensation.

I take this opportunity to place on record my appreciation, as I am sure would many other civilians up and down India, for the contribution to social life and sport made by many many officers of the British and Indian armies. Without their friendly co-operation life could be, and in some cases was, extremely dull.

Simla Sounds

I have heard the breezes rustle
o'er a precipice of pines,
and the half of a *mofussil*
shiver at a jackal's whines.

I have heard the monkeys strafing
ere the dawn begins to glow,
and the long-tailed *langur* laughing
as he lopes along the snow.

I have heard the rickshaw varlets
clear the road with raucous cries—
coolies clad in greens or scarlets
as a mistress may devise.

Well I know the tittle-cattle
of the caustic muleteers,
and the Simla seismic rattle
is familiar to my ears.

Mofussil provincial area, as opposed to presidency enclave, in a district in the rural area   *langur* black-faced, long-tailed monkey with thick grey coat, found mainly in the Himalayan foothills

Though today my feet are climbing
bleaker heights and harder roads,
still the Christ Church bells are chiming,
still the midday gun explodes.

But the sound which echoes loudest
is the sound I never knew
till I launched (the very proudest)
with the staff at AHQ.

'Twas a scene of peace and plenty,
plates a-steam and spoons a-swoop.
'Twas a sound of five and twenty
hungry Generals drinking soup.

J.M. Symns, Indian Educational Service,
      Songs of a Desert Optimist, 1913

A Code of Morals

Now Jones had left his new-wed bride to keep the house in order,
and hied away to the Hurram Hills, above the Afghan border,
to sit on a rock with a heliograph, but ere he left he taught
his wife the working of the code that sets the miles at nought.

And Love had made him very sage, as Nature made her fair;
so Cupid and Apollo linked, per heliograph, the pair.
At dawn, across the Hurram Hills, he flashed her counsel wise—
at e'en, the dying sunset bore her husband's homilies.

He warned her 'gainst seductive youths in scarlet clad and gold,
as much as 'gainst the blandishments paternal of the old;
but kept his gravest warnings for (hereby the ditty hangs)
that snowy-haired Lothario, Lieutenant-General Bangs.

'Twas General Bangs with Aide and Staff that tittupped on the way,
when they beheld a heliograph tempestuously at play.
They thought of border risings and of stations sacked and burnt—
so stopped to take the message—and this is what they learnt:

'Dash dot dot, dot, dot dash, dot dash dot'—twice the General swore.
Was ever General Officer addressed as 'dear' before?
'My love,' i' faith! 'My duck,' Gadzooks! 'My darling popsy-wop!'
'Spirit of great Lord Wolseley, who is on that mountain-top?'

The artless Aide-de-camp was mute; the gilded Staff were still,
as dumb with pent-up mirth, they booked the message from the hill;
for clear as summer lightning-flare, the husband's warning ran:
'Don't dance or ride with General Bangs—a most immoral man.'
With damnatory dot and dash he heliographed his wife
some interesting details of the General's private life.

The artless Aide-de-camp was mute, the shining Staff were still,
and red and ever redder grew the General's shaven gill.
And this was what he said at last (his feelings matter not):
'I think we've tapped a private line. Hi! Threes about there! Trot!'

All honour unto Bangs, for ne'er did Jones thereafter know
by word or act official who read off that helio;
but the tale is on the Frontier, and from Michni to Mooltan
they know the worthy General as 'that most immoral man'.

Kipling, *Departmental Ditties*, 1886

The Shooting Regulations

(To be hung up in every Barrack Room)

When soldiers go forth shooting
     they shall not be less than four,
with a bugler ever tootling
     and two NCOs—or more.

They shall first receive instructions
     as on parade they stand,
on the heinousness of ructions
     with the natives of the land;
on the sanctity of monkeys
     and of peacocks and of fowls,
of doves and mice and donkeys,
     of elephants and owls.

They will then be marched by sections
    to the Quartermaster's stores,
where the armourer makes inspections
    of the arms and sights and bores.

No guns will be permitted
    that will carry thirty yards,
and the soldiers will be fitted
    with little printed cards.

Their names and age and stations,
    their character and size,
a list of their relations
    and the colour of their eyes.

Then after admonitions
    and the signing of each name,
having mastered the conditions
    they may go in search of game.

In a careful scout-like manner
    with tramping martial tread,
with a flaring crimson banner
    a hundred yards ahead.

If a bird should be detected
    which may seem to be *shikar*,
it first must be inspected
    by the nearest *Tahsildar*.

If he shall give decision
    that the fowl is lawful game,
then, aiming with precision,
    the squad may shoot the same.

And, on the word 'cease firing',
    the bugle call will sound,
and any bird expiring
    may be gathered from the ground.

Shikar permitted or lawful hunting of birds and animals for the purpose of eating, adventure, or as part of immemorial custom   *Tahsildar* a subordinate officer in charge of a *tahsil*, a revenue and civil area of administration

Each rule and regulation
    will be posted in a frame
for the welfare of our nation
    (and, perhaps, too—of the game).

    'SCI', *Civil and Military Gazette*

E. &. O.E.

Two soldiers out shooting in southern India accidentally shot a cultivator, who subsequently died. One morning a deputation, evidently just back from the burning *ghat*, presented a document to the *shikari* which read:

To	Captain Chuckmuck	Dr
To	One bloody murder committed	
	Rupees five only	Rs 5
E. & O.E. Received payment		
	Sd. Ramaswamy Pillay	
	(next of kin to defunct)	

Harry Hobbs, *Indian Dust Devils*

A Thousand a Year (A Phantasy of Long Ago)

From the time that he entered the Army—in the days when the Snider rifle was the chief weapon of offence and defence in the hands of the Indian infantryman—Colonel Cantilever, of the 345th Bombay Bukhstiks, had always been a rebel. Not a socialist, or a communist, or any other sort of 'ist', but an habitual disturber of official peace, a chronic kicker against the pricks of babuism. All through his service he had longed to bring home to the authorities the utter futility of many of those moth-eaten anachronisms known as 'Regulations for the Army (India)'.

    Many a memorandum—official, demi-official, and sometimes almost insubordinate—had the gallant and far-seeing Colonel penned to 'higher authority'—that mysterious arbiter of official destinies—in a loyal endeavour to get wrongs righted and matters generally brought up to date and more in keeping with the requirements of the day; but

---

Burning *ghat* steps on a river front assigned for cremation   *shikari* a skilled hunter who hires his services for a day, a week, or a whole shooting season

each and all had been turned down, either by a soft answer ('the matter is receiving attention'), or, after a protracted interchange of acrimonious letters by that official rap on the knuckles—'this correspondence will now cease'.

'If only I had a thousand a year!' he used to say 'I'd show 'em! I'd show up their damned antiquated methods, and bring them to heel, the scoundrels!' The method he proposed, he went on to explain, would partake of some absurd behaviour on his part, some ridiculous interpretation of the 'regulations', an interpretation which, while obviously stretching those poor absurd rulings almost to breaking point, would still however be admissible, and therefore not liable to disciplinary action. But it would require careful management; and in case he overstepped the mark, and was removed from the service for insubordinate conduct, possibly without a pension, it was essential to have something to fall back on.

Years of faithful and gallant service under these soul-destroying conditions had tempered the Colonel's youthful ardour and enthusiasm for reform; and, bowing to the inevitable, he was finishing his last years of regimental command in the old cantonment of Rumbletumble in northern India. The Rumbletumble Brigade was commanded by one of those dear old Generals, whose chief claim to fame lay in the fact that they provided unlimited 'copy' for the facile pens of the more frivolous poets, and served as models for the cruel satires perpetrated on the army officer in the early days of the volunteer movement.

Red-faced and rotund, with a fierce grey moustache, he puffed and blew about the station on a wheezy old horse, swore at his servants and occasionally cantered pompously about the *maidan* amongst his sweating troops, covering them with dust and making bombastic and futile comments on the training of the Army of the day. The evenings generally found him playing golf with a subaltern's wife, or taking the station beauty for a ride; after which he would adjourn to the Club, playing whist with the more ambitious members of his brigade, and then retire to his quarters to dine in solitary state.

And if ever there was a stickler for 'regulations', it was the officer commanding the Rumbletumble Brigade! The bulky volumes containing them were a veritable Bible to him, and it was whispered that

---

Maidan open space in the middle of civil and military stations, used for ceremonial parades, gymkhanas, training forces and sometimes for golf

they reposed under his pillow at night. Colonel Cantilever, of course, regarded him with unmitigated displeasure. The smouldering fires of reform, burnt low after years of fruitless endeavour, were frequently fanned to flame in his breast when dealing with 'this prehistoric old poopstick' (to use the Colonel's own words). But alas! there was no thousand a year to back his schemes, even if he had the desire to carry them out; and after all it was perhaps better, he thought, to conserve such energies as had not been sapped by the pestilential East for the enjoyment of his declining years in the pleasant and evergreen West.

The Rumbletumble Club was gay with coloured lights and bunting. The ballroom was crowded, and not the least energetic among the many dancers was the wicked old General himself. This was long before the days of two-steps and fox trots, and it was to the strains of a popular waltz refrain that the couples glided sedately round the room. It was the occasion of the Bachelors Ball, an annual affair which the whole station attended. Colonel Cantilever stood in the doorway of the ballroom, watching the dancers. He was a non-dancer himself, and never attended these functions except on special occasions like the present. They bored him, and he was now only waiting his chance to slip off to bed. The English mail was due in that night, and, though no delivery would normally be made till the following morning, those who sent a servant down to the post-office could get their mail by means of a 'window delivery'. This he had done and his letters should be waiting in his bungalow now. But he could hardly leave yet; it was too early, and the conventions must be respected. Tired of watching the dancers, he drifted to the bar. There he took unto himself a whisky and soda, a large cigar, and the *Field* and seating himself in the most comfortable chair he could find, he prepared to pass a pleasant hour or so before he said goodnight to his genial hosts.

At length, his drink finished and his cigar half consumed, he rose, and having found his senior host bade him goodnight and walked back to his bungalow. Ah! the mail had come. Good, a letter from his wife, one from his son at school, a couple of catalogues and a long uninteresting-looking envelope. He read his home letters through twice, put the catalogues unopened into the fire, and was just going to do the same with the long envelope when something prompted him

---

The *Field* English magazine devoted to field sports, country houses, and gardens

to open it. He drew from it a neatly folded piece of paper, bearing a London address and headed 'Messrs Barker, Barker and Barker, Solicitors'. This surprised him, but as he read through Messrs Barker's epistle, he grew more and more astonished.

Briefly, it was to inform him that the said Messrs Barker had, on behalf of their late client, Mr John Cantilever, of Perth, Australia, made exhaustive and searching enquiries and satisfied themselves that he, Colonel Cantilever, was the nearest living male relative of their late client, and as such, became their late client's sole heir, 'provided he is in some respectable employ, and bears the same surname as I do, namely Cantilever'—a quotation, evidently, from the dead man's will. Messrs Barker had much pleasure in informing Colonel Basil Fetherstone Cantilever that, though no accurate survey of their late client's affairs had yet been possible, still they were able to assure the aforesaid Colonel Cantilever that, provided his identity was established to their satisfaction, he was undoubtedly the possessor of an income of at least a thousand pounds per annum; and they rounded off this pompous legal document by congratulating him on his good fortune.

Fifteen—even ten—years ago this communication would have sent the Colonel cavorting down the Mall in his pyjamas. Even now it sent a thrill down his spine and made him sit bolt upright in his chair. The shock and surprise served only to rouse his mental activity. The cells of his brain wherein reposed those long-discarded plans were galvanized into action, but from his staid and unhurried movements no one would have suspected that the grey-headed old Colonel was under the influence of an overmastering passion, hitherto dormant, but now roused and stimulated by the receipt of the Solicitors' letter.

He locked the letter away in his writing table drawer, poured himself out a whisky and soda, and, draining it to his reflection in the mirror, went to bed and slept the sleep of the just.

Being naturally a secretive man, he told no one next morning of his good fortune. After all it was his own affair, and he was long past the age when such events send the blood coursing through the veins, in the manner beloved of lady novelists. Besides, he had good reasons of his own for keeping quiet.

After tiffin he told his servant to take a roll of bedding, camp kit, and a small bivouac tent and to pitch the latter near the ninth hole on the Rumbletumble golf links. His servant, used as he was to a great

Colonel Cantilever Camps on Golf Course

deal of strange behaviour on his master's part, stared at him spell-bound and was severely rebuked for his pains.

Behold our hero then, at four o'clock in the afternoon, clad in striped pyjamas and lying on his valise inside a tiny bivouac shelter, pitched as he had ordered, near the ninth hole. Outside stood his servant, gazing towards the ninth tee. Suddenly the servant bent and mumbled something to the prone figure in the tent. The Colonel threw down the book he was reading and rose into as much of a sitting position as the height of the tent would allow. He then deliberately took off his pyjama coat and waited. Presently his ear caught the sounds he was listening for, the gruff tones of the brigade commander and the high-pitched treble of little Mrs Blatherskayte, the newly-married wife of a subaltern. The General was obviously annoyed about something, judging by the tone of his voice. (No wonder! thought the Colonel, but you wait a minute!) When the pair were about ten yards away, the Colonel crawled from his shelter, rose to his feet like some huge animal; (he was as hairy as a bear), and, after a good stretch and a yawn, plunged his face into a canvas basin standing near and began to wash vigorously. For one brief minute the General and his fair partner stood in amazement. Then Mrs Blathers-kayte giggled feebly and turned away blushing; and the General, swallowing his anger with a terrific effort, hurried her away to the Club.

The story was round the station within an hour and many were the speculations as to what action the General would take. The Colonel was known to be eccentric; but surely he must have gone quite potty to behave like that! No one had any explanation; for the rising generation had forgotten, if they ever knew, about the Colonel's wild-cat schemes of many years ago; and Mrs Blatherskayte, temporarily exalted to the position of station heroine, quite lost her head and added details every time she repeated the story. The General main-tained a stolid silence, and, being unable to find anything in his beloved 'regulations' which prevented an officer camping out if he wished to (the spot was just outside the cantonment boundary, curse the fellow!), he was rather at a loss as to what action he should take.

Next day, while in his regimental office, the Colonel received a peremptory note from the brigade major, demanding 'his reasons in writing, for the information of the brigade commander, to account for yesterday's occurrence, which the brigade commander could not believe was either fortuitous or unpremeditated'. The General liked using long words; he thought they added dignity to his communications.

Colonel Cantilever of course was expecting something of the sort, and had his reply ready in his mind. His actions, however, after reading the letter were strange. Taking the envelope in which it had been forwarded, he slit it round the edges and opened it out flat into a sort of irregular polygon. He then took a blunt stub of pencil and wrote a reply on the mutilated envelope. The reasons he gave for his behaviour were briefly that, as the regulations demanded that commanding officers should submit a quarterly return of tentage on regimental charge, including officers' tents, stating whether such tentage was in serviceable condition or otherwise, he had ordered his officers to test their own tents preparatory to the submission of the next return, due very shortly. He was doing so himself when the General saw him yesterday; he was, moreover, extremely sorry if he had interrupted the General's game of golf, but he was unaware that there was anybody on the links at the time.

On receiving this explanation the General fumed. Ringing a small bell on his table, he sent for Brewster, his brigade major. 'Has that damned lunatic Cantilever gone completely off his chump?' he demanded, when Brewster appeared.

'I don't know, sir. Why?' said Brewster.

'Look at that!' thundered the General, thrusting the Colonel's strangely-shaped missive into his hand.

Brewster read it, and his brow wrinkled. He had been vastly amused by the story of yesterday's episode, and truth to tell, he liked old Cantilever very much. But this was a bit too thick! It certainly looked as if the General's surmise as to madness was not far wrong.

'What do you make of it, eh?' asked the General, as Brewster handed back the letter.

'I really don't know, sir. It's difficult to say.'

'Difficult! I should just think it was difficult! Damned nearly impossible, I should say!' roared the General. 'Anyhow I can't be bothered with his lunatic excuses. Just ask him why he can't use decent paper instead of this muck' and he gave the letter a vicious flip.

The receipt of the General's second letter was just what Cantilever had been waiting for. Searching about in the waste-paper basket under his table he extracted a torn and crumpled circular—a bright pink shade—announcing a stock-taking sale of some Calcutta drapery establishment. It was printed on one side only; and, smoothing it out on his blotting-pad, he wrote on the back as follows:

To the Brigade Major
Rumbletumble Bde

With reference to your memorandum of to-day's date, requesting me to furnish my reasons for using this and similar paper for official correspondence, I would point out that, since the Government supply of memorandum forms, as laid down in regulations, is quite inadequate for the requirements of my office, and since I have already used up, in the first half of the year, my total annual allowance of the same, and all requests for a further supply have been refused, I am therefore obliged by circumstances over which I have no control to use such paper etc. as I can obtain free of charge to enable me to carry on.

B.F. Cantilever, Colonel

When this letter (which caused Brewster huge joy) was handed to the General, he was at first inclined to go off the deep end. He was prevented from doing so, however, by a sudden inspiration.

'Tell this half-witted sea-lawyer' he said 'that if he's been so damned careless as to use up his annual supply of stationery in six months, he can damn well pay—pay—d'you hear?—out of his own crazy pocket, if necessary, for his office stationery until the next annual allowance is due. I won't be treated like a damn waste-paper basket, and you can tell him so if you like!'

To this ultimatum the Colonel replied (on the back, I regret to state, of a whisky firm's advertisement circular) that he would exercise more care in future, and would endeavour to make his annual supply of stationery last the allotted period. Meanwhile he was taking steps to supply the deficiency and in future his official correspondence would, he hoped, meet with the General's approval.

Nothing happened for two or three days. The General and Brewster were away for the week-end after duck, and this gave the Colonel the breather he wanted. It was noticed, however, that both he and his adjutant paid an unusual number of visits to the Rumbletumble bazaar.

'Hope that lunatic won't give any more trouble,' said the General when they arrived back from their shoot. 'Damned lucky he's going away, or I'd have to do something about it. When are they off, Brewster?'

'Next week; the 12th I think it is, sir.'

'Oh, as soon as that is it? And a damned good riddance too!'

Totally unexpected was the next development in this little comedy. Brewster, on opening the dak next day, found all the letters from the 345th Bukhstiks beautifully printed on expensive paper, and enclosed in envelopes more suited for a lady's writing desk than a regimental office. He hardly dared take the letters in to the General, and when he eventually did so, the old man thumped the table so hard that the pens and pencils danced in the air, and a huge blot of ink leapt from the ink-pot and landed 'plonk' on his open spectacle case.

'Dammit, Brewster' he shouted 'what the hell's the matter with the fellow? If he wasn't going away I'd damn well put the doctors on to him!'

He flipped through the handful of printed letters (Brewster had wisely burnt the scented envelopes) and then suddenly something seemed to please him; for he leant back in his chair, and looking up at Brewster, said with a chuckle, 'Well, anyhow, this is costing the darned feller something! If he likes to answer letters like this, well let 'im, that's all I've got to say! And I wish his new brigadier joy of him!'

Knowing that it could only last a few days longer, owing to the impending departure of the Bukhstiks for another station, the General was inclined to take a lenient view of the case and to overlook its seriousness. He fondly imagined he had won, and had brought this recalcitrant battalion commander to heel. At any rate he was making him spend some money on all that printing. Damn the feller, though! He was sailing jolly close to the wind, but, search the regulations as he would, he was damned if he could find any that prevented anyone printing official correspondence if they wanted to!

At last the day dawned which marked the departure of the Bukhstiks. All Rumbletumble turned out in its best to bid them good-bye, for they were a popular regiment and everyone—except perhaps the General—genuinely regretted their departure.

But, as their brigade commander, he had of course to see them off officially. He and his staff therefore cantered out that morning to a milestone which they calculated the Bukhstiks would pass at about 8.30 a.m. In the distance they could already hear the blare of the band and could see the dust rising above the trees that hid the marching troops. When the column was about a quarter of a mile away, a sweating orderly ran up to Brewster and handed him a note. It was a big envelope with the Bukhstiks' crest on the back of it. Brewster couldn't imagine what could be inside and tore it open eagerly. What he withdrew from it made him gasp with surprise. It was a beautiful bit of notepaper, called commercially 'Vellum Wove' or some such fancy appellation. Thick as cardboard almost, its glossy surface reflected the rays of the early morning sun sufficiently to make Brewster screw up his eyes as he read. In the top left-hand corner was an embossed and coloured crest consisting of a lemon tree whose branches bore several superb specimens of luscious yellow fruit. On the paper was printed the following letter:

Dear Brewster,

In case the General should ask any questions, I would like to point out that, as the transport allowed by regulations is quite insufficient, several of us, myself included, have been forced to hire extra transport locally, or make other arrangements for the carriage of such of our moderate and perfectly reasonable amount of personal kit for which room cannot be found on the meagre transport allowed. Wishing you and the General the best of luck, and please tell him how sorry I am to be leaving.

Yours sincerely, B.F. Cantilever

Brewster read through this epistle, folded it, and put it in his pocket. He daren't let the General see it; not yet at any rate.

Meantime the head of the Bukhstiks was approaching, and was indeed only a few hundred yards away. A bend in the road hid them from sight, but the noise of the band was plainly audible above the cheers of the spectators.

'Damn funny noise the band's making, isn't it?' said the General. Brewster was no musician, and seldom knew if a band was playing well or not. He was, however, just about to make some banal comment, when the head of the column swung round the bend, and into full view of the General and his staff.

And what they saw made their jaws drop and they stared open-mouthed at the apparition.

For, in front of the band, half of which was incapacitated by hysterics, while the other half gallantly but unmusically stuck to its job, marched Colonel Cantilever. On his head he wore his uniform helmet, surmounted by a disreputable Cawnpore topi. Round his neck was slung a pair of boots; and in front of him he pushed a dilapidated perambulator, full of a miscellaneous assortment of boots, shoes, ties, collars, tennis rackets and polo sticks. On one of the latter was perched, inverted and insecure, an article of domestic utility to which is not usually accorded the honour of marching at the head of a battalion of infantry. As the column approached, Brewster, who had recovered from his first shock and was now verging on hysterics, turned to see the effect on the General.

That gallant old warrior was standing in his stirrups, and his eye-balls looked as if they might drop from his head any minute. His face turned from pink to puce and from puce to purple; then, without a word, he turned his horse and galloped away in a cloud of dust, leaving his staff to watch the last act in the comedy and to see Colonel Cantilever march out with the honours of war.

Lt. Col. E.R.P. Berryman

JUSTICE

EDITOR'S INTRODUCTION

The administration of justice was the last major subject to be tidied up by the British, following the disintegration of the complex and somewhat arbitrary arrangements under the Moghuls. Their first improvisations were not a success and led to still greater chaos. Their final solution, acclaimed by many educated and thoughtful Indians themselves as the greatest, was also one of the most abused of their gifts to India. The principles of English law were not unsuitable because they were alien to the Indian mind, but because, in some respects where this was necessary, they were not separated from the practice of the English courts, which had evolved in very different circumstances. In particular, the mistake was made of removing nearly all administration of justice in rural areas from the scene of the crime, from the village to the magistrates' and judges' courts at headquarters. Justice administered at a distance became justice removed from reality, justice delayed, expensive and open to manipulation. I know this was a controversial question and the introduction of a justice-on-the-spot system would have been bitterly opposed by both Bench and Bar and court staff. Nevertheless, I think the experiment should have been tried. If it had been successful, this would have been bitterly opposed by both Bench and Bar and court staff. Nevertheless, I think the experiment should have been tried. If it had been successful, this would have facilitated the revival of the village *panchayats* to deal with petty crime which constituted 80 per cent of all crime in the rural areas. It must be remembered that the Indian peasant's only recreations were the recurring religious festivals and litigation. The outcome of most cases was uncertain, and a gamble like a bet on the 'Derby' or the 'Grand National'. The item in this section entitled 'The Day's Work' supports my view that much of this litigation was a pastime, a sort of game of one-upmanship. By installing T.V. sets in the villages the Indian Governments hope to reduce the resort to litigation out of sheer boredom.

There were a few provisions of the Indian Code of Criminal Procedure which deserve mention. A confession recorded by the police could not be used as evidence in court. All confessions had to be recorded before a magistrate. This had not eliminated false confessions, but had made it harder for the police to misuse this process of the law. The preventive sections set out in Part IV of the Indian Code of Criminal Procedure were the subject of considerable controversy for some years before Independence. The first-class magistrates were empowered to call for the provision of bonds and sureties from persons likely to cause a breach of the peace or preach or disseminate seditious matter, or vagrants or persons hiding themselves in the district or habitual offenders. Failure to produce the bond or sureties was followed by detention in jail for up to one year, or three years in the case of an habitual offender (section 110). Under section 144, a magistrate could forbid the holding of a public meeting if he was reliably informed that a breach of the peace or some other offence was likely to take place if the meeting were to be held. These were very useful powers for the police and magistracy (rather like the SUS powers of the police in Britain). The Government of Independent India has abolished the preventive sections and, further, has taken all judicial powers away from the Executive. These changes were bound to come but have not necessarily improved the course of justice while weakening the Executive's powers to keep law and order. And there must have been a considerable increase in the numbers of the judiciary to cope with all the work taken away from the Executive.

One way to reduce this extra burden of litigation would be to restore powers to the village panchayats, as I have already mentioned, and, after selection democratically by the villagers, to train the panches and then send them all the petty cases for disposal. One boon arising from such an arrangement would be the almost complete absence of lawyers from the village courts. In the time of the British there was a district and Sessions Judge in every district and in the larger or more important districts, one or even two Additional Sessions Judges. They heard all cases at headquarters.

Perhaps some comment is called for on the principle raised in the contribution called 'Heresy on High'. In England it would be accepted that Judge and Jury should regard with the gravest suspicion the statements of witnesses which were found by cross-examination to be false or unreliable on ancillary or circumstantial points. But in India the reasons for concealing the truth on subsidiary matters might have no relation whatever to the truth on the main issue. Not being well-educated and having a healthy suspicion of all lawyers, witnesses were afraid of making some admission which might involve them in trouble. Such motives cannot be guessed by a newcomer to Indian courts but can sometimes be detected after many years' experience.

On the subject of corruption, Harry Hobbs has described an eighteenth-century British Chief Presidency Magistrate in Calcutta. His practice was crude and likely to gain him the maximum number of dissatisfied litigants in addition to setting a very bad example to the subordinate courts. I like to contrast this with the practice of a well-educated and respectable Judge in a native state. He took 'gifts' from both parties and after he had pronounced judgment in the case, he returned his gift to the unsuccessful party. In a country where gifts of various kinds on various occasions were traditional and part of the daily pattern of life, most Indians who made a practice of calling on Government officers knew that the Government had forbidden its officers to accept any gifts other than fruit, vegetables and nuts. But some still believed that officers, or at least some officers, would be glad to receive more expensive gifts provided that they were conveyed secretly; hence the not uncommon bottle of whisky hidden well below the fruit and vegetables. I did hear of one man who seemingly knew nothing about the prohibition. He presented his Collector with a very fine fish, which was returned with polite regrets. A week later another fish arrived accompanied by a parcel. On opening the parcel the recipient found a canteen of silver fish knives and forks and a note apologizing for not having sent the canteen on the first occasion as a result of which the Collector had been obliged to return the first fish.

Lawyers were often rather a pain in the neck. Of course there were always some very able men, but it was rare to find a lawyer who could argue his case briefly and concisely. The best men naturally went to practise in the High Courts. At the very bottom of the ladder were lawyers who only spoke their own version of English which came to be known as 'Babu' English. We should not have been so hard on them if we had stopped to think what it would have been like for our lawyers if England had been invaded and occupied by, say, Russia, and they had been required to address the courts in Russian. The profession was very popular—not surprisingly in view of the limited openings for able Indians in other spheres of activity.

In British times the heaviest brunt of case-work fell on the comparatively young and junior Indian Civil Service officers and on the older and more experienced officers of the Provincial Service serving as deputy collectors in the districts. Most of us felt like T.F. Bignold when he wrote (p. 250):

Oh, honoured Yule! I would I were like thee,
dispensing justice 'neath a sheltering tree:
and guided less by training than by tact,
could pounce unerring on the trail of fact,
for in those days—'tis long ago, my friend—
Law was the means and Justice was the end.

*

Justice and Mercy

The Chief Presidency Magistrate, a stodgy old-timer otherwise a European gentleman who was somewhat notorious for taking bribes, from both sides, and even then giving a wrong decision. This did not prevent him claiming to be always guided by his conscience. In some ways he resembled that Oriental pillar of justice who, whenever he took his seat in court, piously ejaculated 'We will now proceed to tamper with justice and mercy.'

Harry Hobbs, *John Barleycorn Bahadur*, 1943

On Staying Too Long in India

We heard yesterday of Mr C's death from fever at Agra. It must have been very sudden as no-one had heard of his being ill. He is another victim to the weakness men have for staying on too long in India. His health was not good and he should have retired when he had served his time. He hoped, of course, to get onto the High Court. He has now gone to the Highest Court, but not as Judge.

Mrs R. Moss-King, *Diary of a Civilian's Wife in India*, 1884

The Day's Work

The sun was on the horizon,
the air was filled with the sound
of bells, as the village cattle
returned from the grazing ground.

Nago and Motu and Laxman,
Rama and Ragoji
sat in contemplative silence
under the *pipal*-tree.

Nago had been as a witness
to the courts at Paisapur,
and perjured his soul for a neighbour
who had done it for him before.

Motu had been assisting
a friend in a field dispute
with a rival who had established
his right in a civil suit.

Laxman had gone to the Forest
reserved by the *Sarkar*
and cut some logs and sold them
in the neighbouring bazaar.

Pipal a kind of fig tree, very common in Indian villages   *sarkar* (*sircar*) Government, or its representative; in Bengal, head of affairs (business); denotes administrative division in the Deccan

Rama had been to the *thana*
and made a false report,
to get his enemy Ragoji
into a Criminal Court.

And Ragoji had impounded
a couple of Rama's cows,
driving them off from where they stood
in front of Rama's house.

But now in the hush of evening,
when day's distractions cease,
they sat beneath the pipal-tree,
smoking the pipe of peace.

'G'

Heresy on High

The Bar was boiling over,
the Judge was seeing red:
the latest High Court Ruling
had brought things to a head.

At first they spoke in whispers;
the Judge's head was bowed;
a junior Counsel sniggered,
a senior spoke aloud:

'Your Honour must be knowing
this Ruling which has caused. . . .'
His Honour looked up slowly;
the senior Counsel paused,

the senior Counsel wilted,
the Judge's eye was grim;
it might be better after all
to leave the floor to him.

There followed silence in the Court,
a silence as of death,
what time a lawyer holds his tongue
to take a legal breath.

Thana police station

'I know' his Honour thundered:
the senior Counsel bowed
and took his seat without a word;
the Sessions Court was cowed.

His judgment had been over-ruled,
because the man he'd tried
had been condemned on evidence
of witnesses who'd lied.

'The learned Sessions Judge has erred . . .
We learn with great surprise. . . .'
'The trouble is they never learn'
he thought—and damned their eyes.

Far from the madding benches,
so eloquently dumb,
the Judge's mind was ranging
a past they could not plumb.

He thought of twenty summers' toil
beneath those brazen skies;
of twenty summers lost and gone
in classifying lies.

He thought with unassuming pride,
without a single pang,
on what a thin and twisted line
he'd sentenced men to hang.

He thought of all the pits in which
a Sessions Judge could fall,
and how a High Court Judge could leap
immune into them all.

R.V. Vernède, ICS, Meerut, 1935

[Written on the famous ruling of Mr Justice, later Sir John Douglas, Young, High Court Judge, Allahabad, in respect of a judgment by R.L. Yorke, Sessions Judge, Meerut. Mr Justice Young held that no reliance at all should be placed on the evidence of witnesses who had already been found out doctoring the truth in any part of their evidence. If this dictum were to be accepted without qualification

there would be very few convictions in India! It is related that the
appellant in this case produced a medical certificate to show he was
ill in bed at the time of the alleged robbery with murder and was
brought into the High Court on a charpoy to impress the Appellate
Court of his frailty. Mr Justice Young was convinced that he had been
physically incapable of committing the alleged violence. As soon as
his judgment was announced, the appellant literally took up his bed
and walked out of the court!]

Shi'ir-i-Nazrat (a Palm Song)

Suleiman Khan was a *zubberdust* man,
but fond of the ladies too.
He'd an iron fist and a very long list
of all the villains he knew.

They came to no harm if they greased his palm,
as sensible rascals did;
but no 'Pro Quo' if a fellow said 'No'
and failed to produce his 'Quid'.

A crock of gold to the *daku* bold,
dug up from the *bania's* floor;
a golden fleece for the Chief of Police—
for that was the old *Dastur*.

So all was peace in the land of grease,
until, on a fateful day,
a thief, Dharam Jit, decided to cheat
by hiding his swag away.

There was no *chalan* by Suleiman Khan,
no summons came to obey;
but a strange disease affected his knees
and turned his tegument grey.

He asked his wife and the light of his life
to mix him a strong juleep,
which gave him a pain that affected his brain
so much that he could not sleep.

Zubberdust overbearing *daku* robber *bania* merchant *dastur* custom
chalan equivalent of a police 'charge' in England

He summoned a *vaid* to come to his aid,
who tapped on his swollen tum;
but the medico failed to find why he ailed,
though he charged him a tidy sum.

He invited a priest and gave him a feast—
two goats and a can of *ghee*;
the *Pundit* was kind but quickly divined
that he needed to double his fee.

So help he besought as a last resort
from the men of his own *bhaibandi*;
they met his request, in their Sunday best,
on a suitable date—Palm Sunday.

Silent they sat on a cocoanut mat,
passing the hubble-bubble,
while poor Dharam Jit, as white as a sheet,
told them about his trouble.

'My brothers' he wailed 'I've been assailed
by a most malignant *bhoot*;
please help me repel this fiend from Hell
and you can share in my loot.'

They looked at him then—those honest men—
and their look was a look of greed;
just how much cash could they squeeze as 'dash'
out of this brother in need?

The bargain was made, then each took a spade
and dug up the hidden hoard.
They collared the lot, all Dharam Jit got—
the sack and a piece of cord.

In secret session they parted possession,
each of a fifth of his lot,
to Suleiman Khan in a ruined barn,
a very secluded spot.

Vaid homeopathic doctor   *ghee* clarified butter   *pundit* priest   *bhai-bandi* brotherhood, guild   *hubble-bubble* hookah   *bhoot* evil sprit

the miserable thief was crazy with grief;
he climbed up a handy tree
in a mango *tope* and, taking a rope,
committed 'Felo-de-se'.

What of the moral to draw from this quarrel,
and who was the one to blame?
'Make sure that your wife and the light of your life
is a wholly respectable dame.'

Idem, 1956

Tope grove

2

The Station, Social Setting and Climate

The word 'station' was universally used in India to describe the settlements of British and other European residents in the smaller civil administrative and military garrison centres in India. The station was quite separate from the Indian town or city from which it took its name and on whose outskirts it was situated. Large cities like Calcutta, Bombay, Madras, Delhi, Karachi, were not called stations and the social life there was more closely related to the expatriate life in some European cosmopolitan city. The word 'station' was also used to describe the aggregate society of such a place, e.g. 'The whole station turned out'. British rule at its most meaningful, i.e. at district level, was carried out by its civil officers who lived in and toured from these stations.

Captain George Atkinson, in his book *Curry and Rice*, has described 'Our station', and the pattern was much the same everywhere, with some local variations. Coming from the direction of the Indian town or city, you came first on the Cantonments where the military officers lived in their bungalows and troops in their barracks. Next came the Civil Lines, often very spaciously laid out, in which were situated the club, the Church or churches and the jail. All the Government's civil officers allocated to one district lived in Civil Lines. Some of their bungalows were old and large and dark with an immense thatched roof ending only five feet above the verandah. They were designed for one purpose only—to provide the maximum possible protection from the heat of the sun, and nothing better was devised until the advent of air-conditioning at the very end of British rule. They were not very pleasant to live in except during the hot weather. These older bungalows were set at one end of a spacious compound. A large garden surrounded the bungalow and to one side of this were the stables and the servants' quarters. The rest of the compound was a barren desert, with a few trees perhaps, through which Indians wandered on their lawful or unlawful occasions. If the station was also the headquarters of the district, there would be Government offices, a Treasury, magistrates' and judges' courts, Government schools for Indian children, a rest-house for travellers and a small hospital or dispensary. If it was an important railway centre, there would be a Railway Colony and a Railway Institute.

Captain Atkinson makes no mention of any club, but only of assembly-rooms, where they held dances and amateur theatricals. He also mentions a station swimming bath, which women were not allowed to use. Among station activities which, regrettably, faded away on the advent of motor-cars and more active sports such as tennis, was the Cool of the Evening parade when many residents turned out to 'eat the evening air' by riding down the Mall, the gentlemen on horseback and the ladies in their carriages, all to meet up at the bandstand, where the band played so mightily that a certain amount of flirting between the subalterns and the station's spinsters could be carried out without detection. Presumably all other social activities were carried on in private houses. Atkinson did not mention the club because, with the exception of the Bengal Club in Calcutta (founded in 1827) there were no clubs in India until the numbers of Britons had increased to such an extent that some sort of social centre became essential. I have not been able to ascertain the date of the first station club, but I should be surprised if it was much before 1869. In that year the Suez Canal was opened and with it the floodgates for visitors to India. In *The Chronicles of Budgepore* written in 1870 the author, I.T. Pritchard, makes no mention of a club. Whatever the precise date, the idea spread very rapidly and by the turn of the century there was no station of any size which did not have its own club. The up-country club was quite a humble building, generally in the form of a rather large and rambling bungalow.

The club provided the following facilities: a verandah with chairs, a card room, a billiard room, a bar, a sitting room which might also house the library (books mostly donated by members), a dining room, a ladies' room, changing rooms and a kitchen, together with uniformed staff, and towards the end of the nineteenth-century badminton and lawn tennis courts with a Marker and ball boys. Badminton was actually played under the Poona rules (1876) before it caught on in England. Later some clubs added covered swimming baths, and a very few squash courts. Concerts, dances and theatricals were held in the club unless there was a larger and more suitable room available. This might be the old assembly-room of Captain Atkinson's time or a room in his Town House kindly lent by the local Raja or Nawab. The club was open at most times but only came to life for swims and drinks before lunch; then again for tennis and tea, drinks after tennis and before supper. Billiards and bridge went on before and after supper. Some stayed to supper in the club, but most people went home to have a hot bath and supper at home after changing into evening clothes. Sometimes they went back to the club after supper. The mess, of course, was well patronized by military officers who occasionally invited civil officers to dine as guests in their mess. But, as women were not allowed in the mess, married officers and their wives, and bachelors who liked female company (most of them), flocked to the club which thus became the focus of almost all station life. Not all those who might be eligible actually joined the club, some because they could not afford it, others because they did not want to join.

This is not the place to enter into lengthy polemics for or against not allowing Indians to join the European clubs, except to say that the opposition was not so much because they were Indians but because of their inferior rank and status, i.e. for snobbish rather than racist reasons. When in the last 30 years of British rule, the All-India Covenanted Services began to be Indianized, all station clubs willingly accepted as members Indians and their families who were of equal status and rank to Europeans. In practice it was only a minority of Indians and their wives who joined the clubs, because it was only a few who enjoyed the English style of social life and recreation. All other Indians, officials and non-officials, were normally barred, with one exception—a locally based Raja or Nawab, if he wanted to join. Sometimes a Raja or Nawab, if he did not really want to use the club, would join as a gesture of goodwill but not actually use the club (this reinforces my proposition about snobbery). Provincial-service Indian, Eurasian and Domiciled European officers could probably all join the Indian club, very often run by the lawyers, where they could get bridge as good and tennis of a higher standard than anything which the British club could show.

To return to the subject of snobbery in the station, this was so ridiculously exaggerated as to become a disease. Their bible was the Government 'warrant of Precedence' (15 pages of the Civil List). It lent itself to caricature. The young newcomer, free of such inhibitions, was powerless to influence his elders, just because he was so junior. It is only fair to say that there were always some men and, even more creditably, some women, who were free of these prejudices. Musical or other artistic talent was not recognized as a reason for preferential treatment. The defence for such neglect, though I never heard it mentioned, might well have been that artistic merit was so rare in Indian stations that they would not know how to rank it for social occasions (it was not mentioned in the Warrant of Precedence) and therefore it was better to ignore it.

As a by-product of this snobbery there was a certain type of very senior lady called, briefly but irreverently, a *Burra Mem*. The 'Grand Lady' was always married, for her husband's seniority was her power base. One in a station was not so bad, but a covey of Burra Mems was almost too much to bear. They were the would-be arbiters of fashions and manners, the dictators of morals, the censors of books and plays, the engineers of engagements. Young men who had not been warned in time, or who had not enough *savoir faire* to invent a good excuse were roped in as their *protégés* and ADCs. To survive this test was really more creditable than passing a departmental examination.

One of the drawbacks of life in a small station was the difficulty of obtaining European-type food and stores, though the position improved after the First World War. The local bazaar could produce small and scraggy chickens, small eggs, potatoes, coarse flour and salt, sugar, buffalo's and/or goat's milk, pulse, gram and goat's meat. India produces a number of green vegetables, some similar to those grown in England. If you wanted English vegetables, you could grow them in your own garden, but the gardener would need more instruction for growing vegetables than for growing flowers. However, the housewife had to send away to Calcutta, Bombay, Madras for tinned butter, jam, marmalade, sauces, coffee, chocolate, ham, bacon, sausages. To

---

Burra mem wife of senior British officer, civil or military

sum up, one could feed well enough, but had to be resigned to do without some old favourites.

The climate was always a factor to be reckoned with in the life of the station. In the south of India, the problem was fairly simple, how to keep fit and condition yourself to live all the year round in a humid temperature of 80° F, the humidity increasing somewhat during the monsoon. In northern India there are three seasons, the cold weather from early October to the middle of March; the hot weather from mid-March to mid-June; the rains from mid-June, or if the monsoon is late, from the end of June, to the end of September. The British crowded into the cold weather with its perfect climate all the social activities they could organize—tennis tournaments, gymkhanas, picnics, dances, theatricals, pony races, and, in some places, a low-handicap polo tournament. One of the highlights was the Annual Flower Show. It made no difference what instructions may have been given by the owners of the gardens, their gardeners had only one idea, to produce the largest and most handsome head of the 'florist's' chrysanthemum. Every year they nursed and protected these plants up to the red-letter day in December. It was well understood, not least by the judges, that although there were blooms of equal merit produced from several gardens, the first prize must always be awarded to the Commissioner, and only after that on merit. By the middle of March it was already hotting up; by the end of March nearly all the women left in the station had gone to the hills.

The club cleared its decks, so to speak, and battened down like a warship going into action. The bar and the swimming bath were well patronized. Work continued in courts and offices but not at any special early summer hour, for that was still sacred to the early morning ride. After work some tennis or two slow training chukkas of polo. Another bath and then either home to finish off some report, or a session at the club doing you know what and discussing plans for a shoot the next weekend. Either one went home rather late from the club to have supper at home, or sometimes one joined up with a few other bachelors or grass widowers to go to the local cinema. This showed mostly Indian films, but now and again a Western spectacular. We took with us a crate of bottled beer. When the lights fused, as they did fairly regularly, the beer tided us over until the electricity was restored. I believe that in the hot weather in northern India the longer one spent out of doors doing something active up to 11 a.m.

and again in the evening, the fitter one kept. If one could not do that and did not have to go to court or office, one skulked indoors between 11 a.m. and 5 p.m. using every known device to counter the heat—*khas-khas* tatties (screens) to cover the doors, matting screens hung along the verandahs, *punkah*s, though one tried to reduce their use to evenings and nights. Instead one could instal a 'thermantidote', a fearsome machine nine feet long, four feet broad and seven feet high, which sat on the verandah outside a window. The window was removed and in its place a grass mat was fixed. A hole was cut in this to take a funnel projecting from the thermantidote. Inside were four fans turned by hand which drove the air into the room. To cool the air two large circular holes were cut in the sides of this infernal machine in which grass mats were fixed and water was dripped from perforated troughs above onto these mats, and the surplus water fell into other troughs below to be used again in the troughs above. Its chief defect was that it required at least six coolies to work it, not to mention reserves. Later this was reduced to two men by making the water cooling automatic. With the coming of electric fans in the last 30 years of British rule, the punkah and the thermantidote went out of use except possibly in very small stations which had not yet obtained electricity. In the heyday of the manually operated devices there was a special servant—the *Abdar*—who looked after all these water-works and the punkah-coolies and was also responsible for the making of ice in pans and its storage in wells. He took his name from the element with which he had most to do, namely *ab* (water). When there was no longer a water job for him to do, his title was bestowed on the head servant at the club and, I believe, also in the mess. In the hot weather one slept out of doors, with the mosquito-net folded over the top of the frame to protect one from the dew which sometimes fell at night. There would be no mosquitos till the monsoon rains. Night attire was just shorts. Now and again one would be caught out by a sandstorm. They came very suddenly without any warning. One just had time to get into the bungalow oneself, but unless there were at least two servants wide awake to rescue the bed, it was swept across the lawn until it met some obstacle strong enough to stop it.

Khas-khas aromatic grass root   *punkah* hand-operated fan made from cloth fixed to a long pole, with a rope or leather thong tied to the pole and taken outside the room to be pulled by punkah-coolies

Perhaps the two groups, at opposite ends of the scale, who most deserved commiseration were the magistrates and judges who sweated it out in the hot and stuffy courtrooms, trying cases all day for several days a week, and the British Other Ranks lying naked on their beds in the barracks with nothing to do save watch the punkahs.

'Momos' complained of Prickly Heat and the absence of any proven remedy. I do not recall any treatment which was hundred per cent effective—germicidal soap was a good palliative, but where most victims made a bad mistake—Momos included—was in taking cold baths, for they were the very worst thing for this irritating and painful affliction. Some victims, in the absence of restraint, scratched the little red pinhead rash so savagely that they tore their skin to ribbons.

Another torment of the hot weather, this time of the mind and not the body, was the *koel*, or brain-fever bird, a distant relative of the cuckoo. Agreeing with our versifiers, I doubt whether the world holds any greater abomination. This abomination consists of a call which rises on a scale somewhat irregularly to a point which requires the final note of the scale, but this never comes. The koel reveals its relationship with the cuckoo by laying its eggs in the crow's nest. Emperor Babar was obviously anxious not to offend Indians, when he wrote: 'It has a kind of song and is the nightingale of Hindustan. It is respected by the natives of Hindustan as much as the nightingale is by us.' And to show that it was not only Indians who were tone-deaf, here is a quotation from the diary of a Dutch traveller in 1790:

Le plaisir que cause la fraicheur dont on jouit sous cette belle verdure est augmenté encore par le gazouillement des oiseaux, et les cris clairs et perçans du Koewil.

(The pleasure which one enjoys from the coolness under this beautiful foliage is further enhanced by the twittering of birds and the sharp and piercing call of the Koel.)

There was another wearisome bird of the hot weather, the green or crimson-breasted barbet, who was given his nickname of Coppersmith by Indians and Europeans alike. His 'Tonk! Tonk!' also resembled the cotton-ginning engines which started up in April as soon as the cotton had been harvested.

While their menfolk gasped and drank heavily down below, the womenfolk took to the hill-stations in their dozens like flocks of birds— to Kashmir, Murree, Simla, Chakrata, Mussoorie, Nainital, Darjeeling,

Shillong, Mahabaleshwar, Bangalore, Deolali, the Nilgiri Hills, Wellington. There they settled in hotels and clubs or rented and shared a house from the beginning of April to at least the third week of June and some to the end of September. The normal balance of the sexes was reversed—there were more women than men, but the imbalance was redressed because (a) some of the older women had not come up to the hills to 'beat it up' and (b) there was a steady trickle of young men, including some husbands, coming up for three or ten or fourteen days' leave to escape temporarily from the heat below. The social activities came to their climax in the 'hill-station week', some time in June, with a fancy-dress dance, the bachelors' ball, a tennis tournament, a farce or light opera and, if the hill-station was also the summer headquarters of the Provincial Governor, a ball at Government House.

When the rains arrived about the third week in June, the hill-stations—at least the Himalayan hill-stations—ceased to be attractive. They became shrouded in damp cold mist and rain with just a few breaks. It was cold and a log fire was essential. The alternatives for women in the hills were: to stick it out until the final descent to the plains in October; to go down in July and stay down; or to go down in July for about one and a half months and then return to the hills till the end of September, about another one and a half months. In this way one escaped the worst of the rains in the hills and the worst of the rains in the plains. So much for the great human transhumance.

Meanwhile, what was happening down below? The break of the monsoon was hailed with joy. Sahibs ran out in their pants to enjoy the first heavy shower. All the doors and windows were thrown open to invite the cool breeze. Grass grew up overnight and the 'grass-cuts' no longer had to trek for miles to gather grass for the ponies. Clubs were cleaned in preparation for the first game of golf. Clothes were moved around and aired to prevent mildew. Very soon hordes of flying insects arrived. We had to have a light to eat by, which meant that we shared our dinner with the insects. At first it would rain for hours every day, then later for only a few hours a day, which enabled one to get out for some exercise. The worst part of the rains was when there was a long break, often for ten days. The heat and the humidity became very trying. This was the time when people got boils and probably felt most run down. I always thought that September was the worst month of the year.

I hope I have made it clear how profoundly the weather affected the whole pattern of life, at least in northern India. Much criticism was directed against the annual transfer of Governments and senior Secretariat Civil Servants to the hills for six months of the year. Despite the cost, I think the move was reasonable and sensible. Put it another way. It did not matter whether the work output of the Government of India and of the Secretariats of Provincial Governments was produced in the plains or the hills. But if the Executive Officers in every Province had left their districts for six months of the year it is unlikely that British rule could have been sustained. Now, if we had had air-conditioning it would have been a different story. I may have over-stressed the unpleasant but even as it was, it was a wonderful life and now it has gone for ever.

*

Captain Sprint's Wager

At Scorcheepore the heat is always something to be felt;
its social ice the only thing that's never known to melt.
The people there, perhaps in order somehow cool to be,
are cool to one another in a singular degree.

Of all its icy people, though, the iciest I met
were Martin Brett, Collector, and his wife Matilda Brett.
To freeze all other people seemed the object of his life;
to aid him in his freezing the main object of his wife.

Now once a year these worthies asked the station to a feed,
whereat the welcome, like the wine, was very iced indeed.
The guests were chill, the host and hostess frigid as the Fates;
it was, in fact, a cold collation, even to the plates.

From one of these quite recently, a certain Captain Sprint
to 'stay away', as he would say, received a lucid hint;
that is, alone of all his set, he got no invite card;
and all the tongues were asking why poor Captain Sprint was barred.

Perhaps it was he'd never deigned to pay Matilda court,
perhaps they'd heard that he had said 'he couldn't stand their sort'.
he had this observation made, and friends will often take
such observations to the parties meant, for friendship's sake.

Besides, he was a cheery youth brim full of pranks and fun,
and Scorcheepore was rather sore at things that he had done.
His jolly laugh, his playful chaff, were not much valued there:
to chaff a live Collector is a serious affair.

'Halloa, my boy' his Colonel said 'what's this they're saying now?
Not asked to Brett's? An insult to the race of Sprint I vow.'
'No, Colonel—I'm not asked, but I'll be there, sir, I'll be bound.'
'Not you, my boy!' 'I will, sir. Come, I'll lay you twenty pound.'

'I'll book it' said the Colonel. 'How you'll do it I can't see;
the feed's tomorrow evening, and—Said Sprint, 'Leave that to me,
I'm going to the dinner, though as yet I've no invite,
and you will hand me over twenty pounds to-morrow night.'

The morrow came. Full grew the spacious rooms of Madame Brett.
The Colonel, entering, scanned the guests, but no—no Sprint as yet;
and when the 'rankest' dame was led to dinner by the host,
he thought 'that silly bet of Sprint's was just a piece of boast.'

They sat: each paired correctly with a mate of fitting rank;
a few, of course, dissatisfied, and looking very blank.
There never was a *burra khana* given yet in Ind
where some at the arrangement of the pairs was not chagrined.

The grace. A silence, very long. Then talking in low tone,
disjointed, dropping, fitful; everybody grim as stone.
The servants, in a row, behind each *Sa'ib*'s or lady's back,
like wall of white surmounted by a coping stone of black.

Behind the Colonel's chair there stood one taller than the rest,
with long white coat and turban, and arms folded on his breast.
He placed the Colonel's soup before him when the feed began,
which made the Colonel say, 'Halloa, wherever is my man?'

'No servant come for Sa'ib tonight'—'Ha! absent. Drunk no doubt!
By Gad, to-morrow, sure as eggs is eggs, I'll kick him out'—
'Pray 'scuse me, Sa'ib; he drunken fellow, bringing caste disgrace.
I told to serving Sa'ib this evening. I belong this place.'

Burra khana grand meal   *sa'ib* sahib

Impudence Indeed!

'Ha! one of Brett's, of course. A civil sort of fellow. Stop!
Confound you, don't remove my soup before I've had a drop!'
Then to his fair next neighbour 'Pardon, ma'am: that nincompoop
was on the point of spiriting away my plate of soup.'

'Please, master, 'scuse me: plenty quick they serving courses here;
'fore master eat his soup they bringing fishes round, I fear.
Two fishes coming—one ee pomfret, other tankey sort.'
With that he calmly filled the Colonel's wine-glass up with port.

'Halloa, what's this?' 'Please, 'scuse me if I doing wrong.
Collector always saying Colonel's liking something strong.'
'Some sherry, fool!' 'I fetching master brandy if he like.'
The Colonel muttered something, and he looked inclined to strike.

The Colonel took some pomfret when came round the fishy course:
the fellow promptly over it spooned out some apple sauce.
'Take this away' the Colonel roared. 'Yes, sar! Please, master 'scuse, .
if master only try, with fish this proper sauce to use.'

'By Gad!' the Colonel muttered 'Brett is playing me some goak!
Bad Form of him, by Jingo! and a bit beyond a joke.
I didn't think 'twas in him. I'll be even with him though
some day, by Gad! I'll take occasion just to let him know.'

The entrées journeyed safely through. He got some Salmi fair,
though on the middle of the plate he found a long black hair.
Such things, however, now and then are fated to appear;
and into his champagne the stupid fellow poured some beer.

'Take this away!' the Colonel said, as to his doom resigned.
'Now go and get a slice of mutton, and some jelly, mind.'
It came. The Colonel choked and coughed, and uttered a
        loud 'damn!'
The meat was good: the jelly in his mouth was strawberry jam!

All eyes upon the Colonel. Brett much shocked, indignant too,
apologies, but awkwardness the rest of dinner through.
No further contretemps overt, except a tilt of cream
all down the Colonel's neck, and on his jacket in a stream.

The dinner done, the Colonel said 'Brett, sorry I can't stay.'
'Oh come and hear some music.' 'Many thanks, but not to-day.'
He strode all savage down the hall, dashed on his forage cap,
and shouting 'Bring my buggy!' waited fuming for his trap.

It came, and perched upon it sat the waiter at the feed.
The Colonel said 'By heav'ns, but this is impudence indeed!
Get out, you brute, or I'll get up and break your dirty head!'
'Please, master 'scuse me, master drive me home' the fellow said.

'I do my best for master here; I wait him dinner time;
I making some mistake for master,—surely that no crime.
If I offending master, I jump down and lick the ground;
but master lose his bet to me and hand me twenty pound!'

'Aliph Cheem' (Major Yeldham), *Lays of Ind*, 1875

The Police-Wallah's Little Dinner

I'm somehow feeling a little bored
    with all their district *gup*;
they're not bad fellows—but thank the Lord,
    the party has broken up!

There's McCaul, the Collector, our biggest gun,
    a capital hand at whist,
and passable company, when he's done
    prosing over 'the List'.

I'm sick to death of his grumbling, though,
    for ever about his luck;
and the story I rather think I know
    of every pig he's stuck.

There's Jones, his clever conceited sub.,
    the 'competition' elect,
a youth into whom I should like to rub
    a liniment of respect;

Gup gossip

an honest lad, though a bit absurd!
    And his diction may be choice,
but I think we should like him more if we heard
    rather less of his voice.

There's Tomkins, our Civil and Sessions Judge,
    a pompous ponderous Beak,
who sneers at McCaul's decisions as fudge—
    We know it's professional pique.

On a point of position he's rather a snob,
    at bottom a kindly man:
show that you think he's the district nob,
    and he'll lend you a hand if he can.

There's little Sharp, the surgeon, in charge
    of the Central Suddur jail:
he's a habit of taking very large
    potions of Bass's ale;

a good little fellow—a first-rate pill—
    zealous beyond the ruck;
you couldn't consult a better, till
    nine o' the night has struck.

I've known him do many a kindly act:
    the little man came out strong
when the cholera broke, and the jail was packed
    with the cholera-smitten throng.

Still, after dinner he's hardly fit
    to tackle a question deep;
we find it better to let him sit
    and sip himself to sleep.

There's the Padre, the Reverend Michael Whine,
    the sorrowfullest of men,
who tells you he's crushed with his children nine,
    and what'll he do with ten?

A circle of worthy folk, indeed,
    each of the five, in his sphere;
but it's heavyish work to have 'em to feed
    more than twice in the year.

Two of them think it a favour quite
    to eat my Michaelmas geese.
Let 'em—perhaps they might be right—
    I'm only in the police—

only a Staff Corps skipper, a drudge,
    on a hundred and fifty a week.
Fancy my asking a Sessions Judge!—
    wasn't it awful cheek?

It's nasty too, my 'competitive' friend,
    to stand your bumptious air:
we shall both go home, I suppose, in the end—
    you won't be bumptious there!

First we had Mulligatawny soup,
    which made us all perspire,
for the cook, that obstinate nincompoop,
    had flavoured it hot as fire.

Next a tremendous fragmentary dish
    of salmon was carried in—
the taste was rather of oil than fish,
    with a palpable touch of tin.

Then, when the salmon was swept away,
    we'd a duckey stew, with peas,
and the principal feature of that entrée
    was its circumambient grease.

Then came the pride of my small farm-yard,
    a magnificent Michaelmas goose.
Heavens! his breast was a trifle hard;
    as for his leg, the deuce!

Last, we'd a curry of ancient fowl;
    in terror a portion I took—
Hot?—I could scarcely suppress a howl—
    Curse that fiend of a cook!

The conversation of course began
    anent the coming monsoon:
plentiful rain, said every man,
    would be a tremendous boon.

Paddy was dying, would fail again;
    raggee would never be ripe;
McCaul was anxious to see the rain,
    and hoped it would bring the snipe.

It started him fair, that brutal bird,
    and he rapidly got to boar;
and many a shaggy one's fate we heard
    we'd all of us heard before.

Then 'Competition' must have his say:
    his talk was somewhat big—
shooting snipe was good in its way,
    and so was sticking a pig.

But India wanted a class of men
    of the intellectual type,
with a mind to study, an eye to ken,
    weightier things than snipe—

thinking more of the busy quill
    than the last-invented gun—
striving a noble role to fill,
    and reporting what they'd done.

Hiccoughed the Doctor—'That'sh yer sort,
    that's the ticket for you—
write a couleur de rozshe report—
    write whatever you do.'

'Once knew a fellow who never stirred
    out of his office chair:
wrote a report. Bigwigs inferred
    he'd been everywhere.'

The Reverend Whine here interposed—
    'Brethren, it's very clear,
the best report that is ever composed
    won't stop things getting dear.

Raggee Course red grain—a staple food in south India

It scarcely becomes me to repine
    but it's sad for family men—
already I'm burdened with children nine—
    what'll I do with ten?'

Said the Sessions Judge—'Upon my life
    it's hard a reply to give;
but why did you go and take a wife,
    not being able to live?'

A general laugh. The Collector stout
    thought it a joke immense;
and the Doctor hiccoughed something about
    'Marry, and blowsh ecpensh!'

Then, there was sherry, and handing round
    of ginger and other fruits;
then, in a silence quite profound,
    the lighting of big cheroots.

Then came cards, and soda-and-b,
    onto the snowy board;
and four of us made a whist partie,
    and the little Doctor snored.

I and the Parson lost a mohur,
    at which the Collector joked;
once he forgot himself and swore—
    but then the Parson revoked.

Then there was brandy-pawnee round,
    and the Parson eat some cake;
and the Doctor snored with a horrible sound
    and choked himself awake.

Lastly, we each the other bored
    with the usual district gup,
and then they departed. Oh thank the Lord
    the party has broken up!

                            Idem

Pawnee (*pani*) water

Thoughts on the Third Rubber

When first, poor innocent, I joined this place,
this district club, came villains, smiling still,
craving to know the measure of my skill
at bridge; and like a fool—nay like the ace
of asses absolute—I hawed and hemmed
and said I 'played a little'. Thus my doom
was sealed and Freedom, shrieking, took her flight.
And here I am, poor innocent, condemned
to pass a lifetime in this weary room—
I wonder will we ever dine tonight?

Again, this eve, I'd reached the outer door,
escaping homewards to my well-earned grub;
again the summons stayed me: 'Make a four—
the Doctor and the Judge are in the club.'
Again I fell to that familiar lure:
'One rubber more and then we'll go and dine.'
We'll go and dine! Poor optimistic wight—
One rubber more! Oh sirs, it will endure
unending till to-morrow's sun doth shine!—
I wonder will we ever dine to-night?

Now folk at home are well and truly fed
and seated at the talkies or the play;
the luckiest have even gone to bed;
the second luckiest have put away
three courses. Not a tavern in the town
serves dinner now—at least they'd think it odd.
Even the Spaniard and the Muscovite,
late feeding peoples, now are sitting down
in Seville or in Nijni-Novgorod—
I wonder will we ever dine to-night?

Hard laboured Hercules; eternal Rome
took time to build; Jason had to wait
ere he could bear the fleece of Colchis home;
not swiftly Jacob won the married state;
long slumbered Rip van Winkle; but I claim

all these were records of fantastic speed,
velocities approaching that of light
compared with the interminable game
that holds us foodless when I want to feed—
I wonder will we ever dine to-night?

Charles Hilton-Brown, ICS,
The Gold and the Grey, 1930–5

The Philistine

The Merediths at Sulya,
a year or two ago,
made a pretty garden
at the Judge's bungalow;
little Mrs Meredith
with her own hand,
out of stark wilderness
fashioned fairyland.

With cannas and with cosmos
marigold and rose,
moonflower and jasmine,
with every flower that grows
in the hot plains of India
that beauty do begrudge,
Sulya was sanctified,
when Meredith was Judge.

But the years rolled onward,
the Merediths went,
and what's a mere garden
to the Local Government?
For they sent us MacAlastair,
a vandal, a hun;
and the goats broke the fence
and the garden's done.

Worthy chap, MacAlastair—
works enough for three,
but he doesn't know a canna
from a christmas tree;

he's sacked the best *mali*,
he's sold half the pots,
and the goats eat what's eatable
and the rest just rots.

Often in the evening
of a dull drab day,
I survey the garden
when MacAlastair's away,
and seeing there the handiwork
of this yahoo,
thank the Lord the Merediths
cannot see it too.

Little Mrs Meredith
with her own hand
made beauty out of ugliness
in this forsaken land:
may she end in Paradise
where gardens always grow . . .
As for MacAlastair—
I know where he'll go.

                                        Ibid.

Prickly Heat

In the symptomatic stage, savage warfare did I wage
'gainst a trifling erubescence on the arm,
for I scratched it night and day till I heard some idiot say
that a little iodine would do no harm.

When it spread to hip and shoulder, then I grew a little bolder
and agreed with all the experts at the club
that germicidal soap was the only certain hope,
used gently in the matutinal tub.

But each little feverish pore became a flaming sore
so I cursed and bathed three hours a day instead,
and I used up quite a crowd o' tins of different coloured powder
and I oiled myself before I went to bed.

Mali gardener

But each day I'm getting worse (which explains this scratchy verse)
so my own advice I'll sell you for a song—
Every nincompoop you meet, has a cure for prickly heat,
and every single one of them is wrong.

'Momos'

April

I've no desire to smoke or drink,
nor will I take a hand at cards;
I have no taste at all, I think,
for badminton or billiards.

I do not wish to see the Sphere,
I saw it only yesterday;
I do not want to talk—I fear
I have no funny things to say.

Poor Mrs White is very plain,
and Colonel Gull is fat and placid—
I think I'll just slip home again
and take a little prussic acid.

Idem

V—Ballade of Unao

Far far away
    the Tharu traces
at break of day
    the tiger's paces:
    the wild boar races
out of the *jhao*
    in other places,
not in Unao.

Kasia has pay,
    Shah. paper chases,
Jhansi TA:
    Almora braces;
    familiar faces

Tharu   Gypsy tribe living in the Terai and Bhabar forests at the Himalayan foothills. They were sometimes used as beaters   *jhao* tamarisk

flock in Lucknow
	to balls and races,
not in Unao.

Elsewhere one may
	buy silks and laces,
yet pay his way,
		here who replaces
		his vanished braces?
Buttons? and how
		purchase bootlaces?
Not in Unao.

Blighting my way
		where'er I face is
change and decay.
		The green oasis
		the desert graces
in lands I trow
		where ruled Amasis,
not in Unao.

My locks display
		unwonted spaces.
Care has made hay
		with all their graces.
		The wreath that laces
the prosperous brow
		Time's thefts effaces,
not in Unao.

My lyre, once gay,
		now rust defaces.
All it can play,
		deep in the bass, is
		Eheu Fugaces.—
'Bearer, *le jao*
		baja!'—its place is
not in Unao.

Bearer   valet or head servant, in northern India; same as 'boy' in Bombay and
Malay   *le jao* take away   *baja* any kind of musical instrument

Envoi

Prince, hear the precis;
    my prayer allow—
    write my hic jaces
    not in Unao.

    A.G. Shirreff, ICS,
Tales of the Sarai, 1918

Calcutta in the Rains

Where music (different from the notes
that warble from Italian throats)
with ceaseless din assails—
where crows by day and frogs by night,
incessant foes of calm delight,
croak their discordant lays.

Where insects settle on your meat,
where scorpions crawl beneath your feet,
and deadly snakes infest;
mosquitos' ceaseless teasing sound
and jackals' direful howl confound
destroy your balmy rest.

Bengal Gazette, 12 August 1780; quoted in
    Harry Hobbs, *John Barleycorn Bahadur*

The Gloom Club Dance

When Britons first took to the hills,
preferring to the heat the chills
of Nainital or Simla's thrills,
(despite the cost of double bills)
some genius, by design or chance,
imported from the land of France
a very gay extravagance—
to wit—the famous Gloom Club Dance,
a very Fancy Dress affair,
demanding nerve and also flair
for just how far 'twas safe to dare
in what the dancers chose to wear.

Each year in June the club revives;
the Undertakers' card arrives;
the bachelors invite the wives
and spinsters too to risk their lives
and reputations in disguise,
to gain perhaps the winning prize—
if not, at least to advertise
their charms to some admiring eyes.
Grass widowers are qualified,
like bachelors, to step inside:
but not so husbands closely tied
to jealous wives—these are denied.

And so the frantic search begins
for new ideas. Inspired by gins,
some opt for woad and some for skins,
and every shop runs out of pins.
The Gloom Club welcomes every guest,
however drunk, however drest,
and hopes run high in every breast
that those who judge will be impressed.
Armed with a stop-watch and a 'scotch',
the Mutes and the Chief Mourner watch
with hooded eyes the strange hotch-potch
of couples shuffling in the 'nautch'.

For every age there is a dress,
from Boadicea to Good Queen Bess,
from *burka* to a bead or less—
for men—well you would have to guess.
Two Adams and as many Eves
dance to a lullaby of the leaves,
while here and there a dusky Jeeves
ignores impassively what he perceives.
While svelte Sumerian ladies stare
with envy at Godiva's hair,
their seniors indignant glare,
for back and sides go bare, go bare.

Nautch dance   *burka* sac-like covering from head to foot, worn by Muslim
women in *purdah*

The Gloom Club Dance

How regally the *burra-mems*
sail round the floor with trailing hems,
like barges on the river Thames,
steered round by tugs with stratagems!
Rank and position we ignore:
the man whom millions hold in awe,
his partner thinks a crashing bore—
no leveller like the dancing floor.
Here stoops to folly once a year,
the High Court Judge, be-wigged, severe;
those lips, which evil-doers fear,
to-night breathe in his partner's ear.

That portly figure touched with grey,
benignant overlord by day,
who rules on postings and on pay,
at night has only feet of clay.
There, flushed with potent wine of France,
the amorous expert in finance
with difficulty keeps his stance,
turning in figures of the dance.
See, massed behind his second front,
the burly master of the hunt
pursues his line, direct and blunt—
his wretched partner bears the brunt.

A charioteer!—surely he's daft?
(but nearly everyone has laughed)
cunningly steers his home-made craft,
two damsels harnessed to the shaft.
Another, sinister and tall,
man of the trees, outdoes them all:
his gleaming leer would appal
the villain of a Music Hall.
Young Joan of Arc (that's her disguise)
lisps sweetly and divinely sighs;
she may not win, but, if you're wise,
be careful of those bedroom eyes.

And, when the seniors have gone,
the fun goes on until the dawn,
with half the fancy dresses torn,
and some are drunk, while one forlorn
seeks hopefully a *kala jagah*
and some young subaltern to hug her,
who's not engaged in playing rugger
all round the dance-floor hugger-mugger.
The dance is over, prizes won,
the guests are leaving one by one.
High in the heavens rides the sun:
another day has just begun.

Game to the last like wounded fauna,
stewed like the bathers in a sauna,
the Mutes, still dumb, and the Chief Mourner
lie prostrate in the judges' corner.

Anon, 1933

Ode to the Brain-fever Bird

Out on thee damned Spirit,
bird thou never wert,
no, nor even near it,
who outpours thy heart
in such maddening strains of syncopated art.

Higher still and higher
in the scale ascending,
like some spinster choir,
frenzied voices blending
at their topmost pitch, all decency transcending.

Shriller yet and shriller
sounds thy piercing note,
as in stucco villa
from suburban throat
grand operatic trills upon the numbed ear float.

Kala jagah dark place or corner sought out by spooning couples

Like a cracked steam whistle,
like an unoiled brake,
like the shriek a thistle
sat on by mistake,
pricks from pompous gent whose hams with fatness ache.

Could I in these verses
let my feelings go,
such ensanguined curses
from my lips would flow,
the reader's brain would reel as mine is doing now.

'Momos'

3
Servants

Servants played a very important part in the life of the British in India. With their assistance the British were able to maintain some semblance of European life-style in reasonable comfort despite a hostile climate and strange customs, e.g. method of shopping. In the course of twenty or more years' service, very close ties of affection grew up between many Indian servants and their British masters. There is no real significance in the fact that many of the portraits of servants in this collection are hardly flattering. The balance is undoubtedly wrong. Most Indian servants were trustworthy, ingenious and honest according to their own code, which allowed for certain small perquisites. Yes, the balance is wrong, but eulogy can become tedious; it is always more amusing to write and read about rogues.

There were two main categories of servants:

1. Those who worked for some British officer or family by day but returned to their homes and families at night. These servants, mostly of humble status, did not accompany their employer when he was transferred to another station, but hoped to be taken on by his successor.

2. The higher-caste or higher-grade personal servants who lived in the servants' quarters inside the compound of their master's bungalow, with, but more often without, their families. They moved with their master whenever he was transferred, unless he took this opportunity to dispense with the services of a servant who was not up to the mark. Some of these men stayed with one master for the whole period of his career in India.

Taking an average-sized staff in northern India in the last twenty years of British rule:

Servants in category 1 were:	Servants in category 2 were:
The sweeper (*mehtar*)	The head servant (*bearer*)
The water-carrier (*bhisti*)	The cook (*khansama, bawarchi*)
The scullion (*masalchi*)	The waiter (*khidmatgar*)
The gardener (*malee*)	The grooms (*syces*)
The garden-coolie and the grass-cuts	The Lady's maid and nanny (*ayah*)
The washerman (*dhobi*)	

The dhobi, who worked for several families, charged piece-rate, or so much per head or per family. The *darzi* (tailor), who once ranked as a family servant in category 1 and almost literally made everything the family wore, ceased to be so much in demand when clothes shops began to flood the bazaars and when British wives were able to get home more frequently and buy the clothes they needed. So now, like the dhobi, the darzi worked for several families and charged piece-rate. All part-time labour, such as punkah-coolies, came under category 1. The night-watchman (*chowkidar*) and orderlies (*chuprassis*) were engaged and paid by the Government and always stayed in the same station.

It is difficult to give a convincing explanation to people who have never lived in India for the large number of servants kept by the British in India. There were several reasons, all peculiar to India. The earliest British residents did not want to offend the natives, so did not cavil when they were told that by customary and religious usage, the higher servants' posts could only be filled by high-caste Hindus; that the more menial tasks could only be performed by lower-caste Hindus, and, in one case—the most menial task of all—that of the sweeper, could only be performed by an untouchable. They were told that the higher-caste or -class servants could not be asked to perform any task beyond that for which they had been engaged, and in no case any task for which a lower-caste man had been engaged. These restrictions only applied to Hindus, but when the Europeans rather naturally tried to engage Mahommedan servants, they found that these too had ordered their own hierarchy and limitation of duties, based on a shrewd observation of the status of each post under offer. So one way or another, the result was a very inflexible system of one man, one job, and almost no interchange of tasks. In the course of time, by the twentieth century, these restrictions came to be somewhat relaxed, especially in the

Hindu lower-middle-caste range and amongst higher-class Muslim servants.

In the very early days, when the British came out, not to rule, but to make money, much larger numbers of servants than could possibly have been needed, were engaged in deliberate ostentation, to impress the Indian merchants and rulers. The greatest ostentation was shown in the number of servants retained for travel and for protection. This phase lasted up to the end of the eighteenth century. In 1782 William Hickey retained 63 servants. Of these eight formed a guard of stick-bearers (*chobdars*) and 20 formed teams for carrying their employer. By the standard of his day this was quite a modest retinue. His extravagance, rather, was a projection of his extravagant way of life in England. He was a rake, determined to eat and drink well and to drive dangerously. His retinue included nine valets, two cooks, two bakers, a hairdresser, a wig-barber, a hookah-bearer, and he must have had the humbler servants as well, sweeper, water-carrier, gardener, etc.

By the turn of the century the East India Company had started to send young men out without any previous training and still called writers but to learn by practice on the spot how to become successful judges, magistrates, collectors. After several changes of policy, it was finally decided that these men should reside out in their districts. Whereas officers under the rank of Captain in Calcutta were required to keep not less than thirty servants, and some kept many more, this new class of Company's men were not allowed to trade and their salaries were not high enough to enable them to engage any more servants than they really needed. Henceforth every British officer and family tried to keep the number and cost of their servants down to the minimum possible. Remember that they needed more servants than their successors in the last thirty years of the Raj. They kept cows and goats and hens, palanquin-bearers for any long journey, a coachman and grooms for four horses for their carriage and for riding, punkah-coolies to enable them to sleep at all in the hot weather.

This brings me to by far the most compelling reason for the engagement of certain key staff, namely, the absence of all basic services and facilities. There was no public transport until the railway network was established in the last thirty years of the nineteenth century; no power, no uncontaminated mains water, no sanitary covered drainage. Provision for private transport was the most expensive. In 1792 Thomas Twining, an enterprising bachelor who travelled extensively, employed forty-four servants for the purpose of travel alone. The roads were bad, the rivers hazardous, and robbers frequently met with. Twining had twenty boatmen and an armed guard of ten. He himself rode in a trotting bullock cart, the best vehicle that India ever produced, as it could go anywhere and was cheap to make, to run and to repair. The railways changed all this.

The only other, and that only a limited, service to be provided was electricity. But this was only provided in the big cities and in the larger stations in the 1920s. The private companies who supplied this, only generated enough power to provide lighting and fans. It was a good thing that punkahs and punkah-coolies could be dispensed with, though the fans were by no means ideal, being too noisy and apt to drive stale hot air round and round the room. The substitution of electric for oil lamps brought no saving in staff as the lamp boy was probably already doing the washing up as well.

In most houses there was a verandah above steps in front, behind and sometimes all round the house. There was a wide front door which remained open all day in the cold weather and rains. On each side of the doorway sat the Orderlies, who announced and showed in both official and private visitors. This door led into a wide passage which ran through the whole bungalow from front to back. All the living and bedrooms opened onto this central passage. All rooms on the same side of the central passage were interconnected by doors. All doorways were high, the doors double and ill-fitting. As a result dust was always blowing in and this kept the sweeper busy. This design was only justified during the rains, when, by opening all the doors, every room in the house could enjoy the cool breezes. Each bedroom had its own dressing-room and bathroom, opening onto the verandah or back of the house. The bathrooms were furnished with a zinc or galvanized iron tub, a large enamel or tin mug, a wooden bath-side platform and an earthenware jar of cold water (*chatti*), all standing in a raised enclosure across one corner of the bath-room, bounded by a raised concrete rib to stop the bathwater running all over the rest of the bathroom. Outside the raised enclosure stood a Victorian washstand with jug and basin. It was the duty of the water-carrier to draw water from the nearest well, to carry it to the house and distribute it to every bathroom, the pantry and the kitchen, and to supply water to the gardener, unless the latter had made other arrangements (e.g. a pair of bullocks using a ramp). Drinking water came from the same source provided it was a sweet-water well—if not, then from the nearest sweet-water well. It was boiled and filtered before use. With the hot weather and winds the bhisti was required to sprinkle water on the drive to lay the dust, and to supply water to coolies temporarily engaged to throw water against the outside of 'tatties', or frames made of an aromatic grass root (*khas-khas*) which converted the hot wind into a delicious cool scented breeze. The bhisti carried his water in a goatskin hide. He took care to protect his lungs and other vital organs by wrapping several folds of cloth or leather between his body and the cold wet skin. Hot water was heated in kerosene tins over an open wood fire and the bath water smelt deliciously of wood smoke.

The only other piece of furniture in the bathroom was a wooden commode. The sweeper, whose unpleasant task it was to empty the contents, developed a kind of extra-sensory perception to tell him when the bathroom was still occupied. His second line of defence was a modest cough just outside the door. This was an infallible warning by which to avoid an embarrassing confrontation. Sweepers were generally small men, going about their work sadly, bowed under huge stickless brooms and always ready to dust the rooms, sweep the floors, the verandah, the steps or the drive, or to clean and feed the dog, without any nonsense of 'one man—one job'. I never enquired or discovered where they deposited, or buried or sold the fruits of their toil, for it was a delicate subject, on which I might learn something to my disadvantage.

There was no kitchen inside any Indian bungalow. There being no source of power, cooking had to be done over coal or charcoal. Apart from the risk of fire, the heat and smell would have made the rest of the bungalow unbearable. There was a small pantry off the dining-room, equipped with a 'hot case' to keep food warm until served up at table. Washing up was also done in this pantry. Best not to enquire too closely how this was done. The kitchen, a small murky room, was normally up to thirty or forty yards behind the bungalow, sometimes with a covered way between. There in mud-brick ovens above charcoal fires, in an atmosphere of choking smoke and reeking spices, the Indian cooks would produce the most marvellous dishes. There were two schools of thought about kitchen inspections. One advised the Memsahib to pay unannounced visits at irregular intervals to keep the cook on his toes and to eliminate the worst hygienic horrors. This

Bhisti and Sweeper attending Victorian Loo

policy was followed by some older, more experienced women and, of course, strongly advocated by Flora Annie Steel in her famous Indian Cookery Book. The other advised the Memsahib to keep well away from the kitchen or risk not only her own demoralization, but, more important, the demoralization of the cook as well. I regret to say that most Memsahibs, especially the younger ones, after a short trial of the former method, came down firmly in favour of the second school of thought.

In the small stations there were no shops as we know them, only bazaars. It could be quite interesting and amusing to go down to the bazaar to buy a length of cloth or some sandals. With your orderly breathing down the shopkeeper's neck, you were not likely to be much overcharged. But it was quite a different kettle of fish when it came to buying food. But why try? Traditionally it was the Indian cook who bought all the food. He would start out early, would know exactly where to go for each item, would bargain and argue with the shopkeeper and, if he could not play off one against the other, would threaten both with the loss of the Sahib's custom. When he had finished, he hired a small boy with a round basket on his head to carry all his purchases back to the bungalow. For this service it was well worth paying the cook his small commission.

The stables—a word may not be out of place about riding, the best and most popular recreation of the British in India. There were stables attached to the servants' quarters of most bungalows. The Government expected every gazetted officer to own, keep or hire a horse (charger). A married officer often kept two horses, which also enabled him to play station polo and if he became good enough, to play in low-handicap polo tournaments. It also enabled him to take part in pigsticking. As you will by now have been led to expect, each horse had its own groom plus a male but more often a female 'grass-cut'. Over nine-tenths of India there was no grazing as we know it at home. There were no boundary fences or hedges, just low earthen banks between fields, and the fields either lay fallow or carried some kind of crop. So the grass-cuts left very early in the morning and came back in the evening with mountainous loads of grass balanced on their heads. Again, I never enquired or discovered where they got their grass; but I never received a complaint. In the rains it was easier as grass sprang up everywhere.

Of all the servants the cook was the most temperamental, but this is the hallmark of cooks the world over. Again, I would regard with suspicion a cook who had no vice. On the other hand, excessive resort to the bottle was the undoing of many a good cook. Neverthe- less, they remained in circulation, for the supply never equalled the demand.

The *ayah* was also apt to be volatile and at one time or another quarrelled with nearly all the male servants. As to her duties, she was first and foremost her Mistress' maid and so in her own eyes the equal of the Sahib's valet (Bearer). The latter, especially if he was also the head servant, naturally did not agree with her, nor did any of the other servants. But they feared her mischievous and tell-tale tongue. So in practice an uneasy truce prevailed. If there were children she would act as nanny. I cannot do better than quote Flora Annie Steel in her *Complete Housekeeper and Cook*:

We may only remark that, with very few exceptions, Indian ayahs are singularly kind, injudicious, patient and thoughtless, in their care of children: but to expect anything like common sense from them is to lay yourself open to certain disappointment.

A few words for the waiter. He was a middleman, so to speak, on the one hand keeping the kitchen at a respectable distance and on the other guarding us from the horrors of washing up by the scullion. He ruled in the pantry and dining room. His duties were to set the table, warm the plates if necessary, fetch the food from the kitchen (unless he could bully someone else to bring it to the pantry) and put it in the hot case until required to be served up, and of course to serve at table. After the meal he washed up the best dishes and silver, while the *masalchi* washed up the commoner crockery and knives, forks and spoons. The ritual followed in upper or upper-middle-class families in England was almost universally accepted as applicable to India. British wives in India had perhaps their greatest satisfaction in training Indian waiters, so that they became the equals of the best waiters anywhere.

I have said that Indian servants were ingenious. One of their most successful yet commonsense ploys was the device of getting extra waiters for a large dinner party by requisitioning the services of the private waiters of all the families invited to the dinner, and also some of their crockery. This was a well understood operation and it was not done for the families invited to show any surprise or even to recognize their own servants waiting on them, or their own crockery on the table.

With so many servants it was important to try and secure for the head of your staff a sirdar (head) bearer or butler who could command the obedience and respect of the other servants (except possibly the ayah and a strong-minded cook). A firm and impartial head servant would look after the rest, settle their minor quarrels and report anything more serious to his employers. He would satisfy himself of their various needs and, where justified, would include them in his household accounts. Such a man would be some insurance against thefts by any of the staff. He made all the arrangements for travel, bought the railway tickets and brought hot water for shaving and early morning tea to the carriage. He and a minor official called the *nazir*— a sort of quarter master—made all the arrangements for camping. In short, he supervised everything but was not too lordly to mend his master's socks and sew on his buttons.

Some comparative figures of costs may be of interest. They are taken from the United Provinces in central upper India and therefore may not be accurate for every British household in India. But I believe there was a family likeness.

In 1878 Mrs R. Moss-King, the wife of a District Judge in the UP, kept, as noted in her diary, thirty-two servants, of whom twenty-four were basic full-time staff and eight part-time. The following are the details together with the monthly wages:

Servants	Rs per month
Full-time basic	
Khansamah (butler)	10
Bawarchi (cook)	10
Bearer (valet)	10
2 Khidmatgars (waiters)	16
Masalchi (scullion)	5
Ayah (Lady's maid)	10
Dhaie (wet-nurse)	10
Mehtar (sweeper)	5
Mehterani (sweeperess)	4
Bearer's mate	6
Bheesti (water-carrier)	5
Dhobi (washerman)	13
Darzi (tailor)	10
Coachman	8
3 Syces (grooms) @ 5	15

3 Grass-cuts @ 4	12
Malee (gardener)	6
2 Garden-coolies @ 4	8
	163

Part-time basic

6 Punkah-coolies for 5 months @ Rs 4	24
1 Gwala (cowman)	5
1 Murghi-wala (hen-keeper)	5
	34
For 5 months @ Rs 985	
For 7 months @ Rs 1211	
For 12 months	2196

Exchange rate in 1878—the Rupee was worth approximately 2/1d. Therefore, Mrs Moss-King's servants must have cost her approximately £228. 15s a year.

In 1928 an average married family in the United Provinces kept a basic ten servants, or if they kept two horses, as many did, a basic 14 servants. I have included the dhobi as his service was essential, but omitted the darzi as he was only called in occasionally and then paid piece-rates.

Servants	Rs per month
Full-time basic	
Bearer (Valet and Head of Staff)	20
Khansamah or bawarchi (cook)	20
Khidmatgar (waiter)	15
Masalchi (scullion)	6
Mehtar (sweeper)	8
Bhisti (water-carrier)	8
2 Syces (grooms) @ Rs 10	20
2 Grass-cuts @ Rs 8	16
Malee (gardener)	10
Garden-coolie	5
Dhobi (washerman)	15
Ayah (Lady's maid)	20
	163
Total	1956 p.a.

With the rupee at approximately 1/6d. the annual cost of servants' wages only was approximately £146. 14s. But wages were not the only expenses incurred on servants. One was expected to give warm coats in winter to those servants who could be expected to look after them properly, and white uniforms to the same men for the rest of the year. You were expected to provide them with belts and *puggarees* in your own service colours and all railway fares when travelling on duty. The full cost was not much under £200 per annum. As for ability to pay such wages, I take, as an example, the case of a young civilian with nine years' service, married and with one child, and officiating as a district officer. His salary was £120 p.m. His bare servants' wages without other expenses cost approximately £12. 2s. p.m. In annual terms his salary was £1440 and his wage bill approximately £200.

As to the number of servants, there were precedents and standards long accepted which ruled master and servant throughout the subcontinent. It would have been almost impossible to have ignored them.

In Malaysia the British only needed two or three household servants, a Chinese cook and Tamil boy. Because of the absence of caste restrictions three servants could do all the work between them which required twelve servants in India. Moreover, the British in Malaysia were, very sensibly, content with rather more easy-going standards.

I hope that what I have said in this rather lengthy note will help to explain why British families in India found it necessary to engage so many servants. Such numbers were never questioned by Government, nor by any politician, nor by any lobby. Nor were they questioned by any Europeans resident in India. And Indians themselves would have been very shocked if they had found a British family, especially of the ruling class, trying to 'do' for themselves. They would remember, if only by oral tradition, the display which accompanied their own rulers in the past and the many jobs this provided, and would expect their present rulers to keep up something, if only a shadow, of that ancient splendour. The second and more earthly thought would have been to despise this Sahib and Memsahib for their meanness. Which brings me to my final point. In India it was far more sensible to follow tradition, especially when this was more convenient, than to be led into aberration by humanitarian

---

*Puggaree*s turbans

or moral arguments. Whether this was a worthy attitude, I leave philosophers to argue.

With Independence, the whole servant structure of the Raj has disappeared. Continually mounting inflation has meant that only the very wealthy can afford more than two or three servants. So a structure which lasted for nearly two centuries has been blown away in something under twenty years.

*

Engaging a Boy

What a wonderful provision of nature the boy is in this half-hatched civilization of ours, which merely distracts our energies by multiplying our needs and leaves us no better off than we were before we discovered them! He seems to have a natural aptitude for discerning, or even inventing, your wants, and supplies them before you yourself are aware of them! What is the history of the boy? How and where did he originate? What is the derivation of his name? I have heard it traced to the Hindostanee word *bhai*, a brother, but the usual attitude of the Anglo-Indian's mind towards his domestics does not give sufficient support to this. I incline to the belief that the word is of hybrid origin, having its roots in *bhoee*, a bearer, and drawing the tenderer shades of its meaning from the English word which it resembles. To this, no doubt, may be traced in part the master's disposition to regard his boy always as *in statu pupillari* and the boy on the other hand, cheerfully regards him as *in loco parentis* and accepts much from him which he will not endure from a stranger. The boy is often very much a reflection of the master. There are boys and boys. There is a boy with whom, when you get him, you can do nothing but dismiss him, and this is not a loss to him only, but to you, for every dismissal weakens your position. Believe me, the reputation that your service is permanent, like service under the *Sircar*, is worth many rupees a month in India.

The engagement of a first boy, therefore, is a momentous crisis,

---

Boy   head servant in Bombay, equivalent of 'bearer' in northern India and Bihar, and 'butler' in Madras   *sircar* (*sarkar*)   the Government; hence any representative of the same

fraught with fat contentment and a good digestion, or with unrest, distraction, bad temper, and a ruined constitution. Unfortunately, we approach this epoch in a condition of original ignorance. There is not even any guide or handbook on boys. The griffin a week old has to decide for himself between not a dozen specimens, but a dozen types, all strange, and each differing from the other in dress, complexion, manner and even language. As soon as it becomes known that the new Saheb from England is in need of a boy, the levee begins. First you are waited upon by a personage of imposing appearance. His broad and dignified face is ornamented with grey and well-trimmed whiskers. There is no lack of gold thread on his turban, an ample cummerbund envelops his portly figure, and he wears canvas shoes. He left his walking cane at the door. His testimonials are unexceptionable, mostly signed by mess secretaries; and he talks familiarly, in good English, of Members of Council. Everything is most satisfactory, and you enquire timidly, what salary he would expect. He replies that that rests with your lordship: in his last appointment he had Rs 35 a month and a pony to ride to market. The situation is now very embarrassing. It is not only that you feel that you are in the presence of a greater man than yourself, but that you know he feels it. By far the best way out of the difficulty is to accept your relative position, and tell him blandly that when you are a Commissioner Saheb or a Commander-in-Chief, he shall be your head butler. He will understand and retire with a polite assurance that that day is not far distant.

As soon as the result of this interview becomes known, a man of very black complexion offers his services. He has no shoes or cummerbund but his coat is spotlessly white. His certificates are excellent, but signed by persons whom you have not met or heard of. They all speak of him as very hard-working and some say he is honest. His spotless dress will prepossess you if you do not understand it. Its real significance is that he had to go to the dhobi to fit himself for coming into your presence. This man's expectations as regards salary are most modest, and you are in much danger of engaging him, unless the Hotel Butler takes an opportunity of warning you earnestly that 'this man not gentlyman's servant, sir! He sojer's servant.'

The next who offers himself will probably be of the Goanese variety. He comes in a black coat with continuations of checked jail

---

Sojer   soldier

cloth, and takes his hat off just before he enters the gate. He is said to be a Colonel in the Goa Militia, but it is impossible to guess his rank, as he always wears mufti in Bombay. He calls himself plain Mr Querobino Floriano de Braganza. His testimonials are excellent; several of them say he is a good tailor, which to a bachelor is a recommendation; and his expectations as regards his stipend are not immoderate. The only suspicious thing is that his services have been dispensed with on several occasions very suddenly without apparent reason. He sheds no light on this circumstance when you question him, but closer scrutiny of his certificates will reveal the fact that the convivial season of Christmas has a certain fatality for him.

When he retires, you may have a call from a fine looking old follower of the Prophet. He is dressed in spotless white, with a white turban and white cummerbund; his beard would be as white as either, if he had not dyed it a rich orange. He also has lost his place very suddenly more times than once and, on the last occasion, without a certificate. When you ask him the cause of this, he explains, with a certain brief dignity, in good Hindostanee, that there was some *tukrar* (disagreement) between him and one of the other servants in which his master took the part of the other, and as his *abroo* (honour) was concerned, he resigned. He does not tell you that the tukrar in question culminated in his pursuing the cook round the compound with a carving knife in his hand, after which he burst into the presence of the lady of the house, gesticulating with the same weapon, and informed her, in a heated manner, that he was quite prepared to cut the throats of all the servants if honour required it.

If none of the preceding please you, there will be the inevitable unfortunate whose house was burnt to ashes two months ago, on which occasion he lost everything he had, including, of course, all his valuable certificates. Another will send in a budget dating from the troubled times of the mutiny. From them it will appear that he has served in almost every capacity and can turn his hand to anything, is especially good with children, cooks well, and knows English thoroughly, having been twice to England with his master. When this desirable man is summoned into your presence, you cannot help being startled to find how lightly age sits on him; he looks like twenty-five. As for his knowledge of English, it must be latent, for he always falls back upon his own vernacular for purposes of conversation. You rashly charge him with having stolen his certificates, but he indi-

A Fine Looking Old Follower of the Prophet

gnantly repels the insinuation. You find a discrepancy, however, in the name and press him still further, whereupon he retires from his first position to the extent of admitting that the papers, though rightly his, were earned by his father. He does not seem to think that this detracts much from their value. Others will come; the larger the series of specimens which you examine, the more difficult it becomes to decide to which of all of them you should commit your happiness. Characters are a snare, for the master when parting with his boy too often pays off arrears of charity in his certificate; and besides, the prudent boy always has his papers read to him and eliminates anything detrimental to his interests. But there must be marks by which, if you were to study them closely, you might distinguish the occult qualities of boys and divide them into genera and orders. The subject only wants its Linnaeus. If ever I gird myself for my magnum opus, I am determined it shall be a 'Compendious Guide to the Classification of Indian Boys'.

E.H. Aitken, *Behind the Bungalow*, 1889

The Boy at Home

Your boy is your *valet de chambre*, your butler, your tailor, your steward and general agent, your interpreter, or oriental translator and your treasurer. On assuming charge of his duties he takes steps, first, in an unobtrusive way, to ascertain the amount of your income, both that he may know the measure of his dignity, and also that he may be able to form an estimate of what you ought to spend. This is a matter with which he feels he is officially concerned. Indeed the arrangement which accords best with his own view of his position and responsibility is that, as you draw your salary each month, you should make it over to him in full. Under this arrangement he has a tendency to grow rich and, as a consequence, portly in his figure and consequential in his bearing, in return for which he will manage all your affairs without allowing you to be worried by the cares of life, supply all your wants, keep you in pocket money and maintain your dignity on all occasions. If you have not a large enough soul to consent to this arrangement, he is not discouraged. He will still be your treasurer, meeting all your petty liabilities out of his own funds and coming to your aid when you find yourself without change. As far as my observations go, this is an infallible mark of a really respectable boy, that he is never without money. At the end of the month he presents you a faithful account of his expenditure. There is a mystery about these accounts which I have never been able to solve. The total is always, on the face of it, monstrous and not to be endured, but when you call your boy up and prepare to discharge the bombshell of your indignation, he merely enquires in an unagitated tone of voice, which item you find fault with, and you become painfully aware that you have not a leg to stand on. In the first place, most of the items are too minute to allow of much retrenchment. You can scarcely make sweeping deductions on such charges as: 'Butons for master's trousers, 9 pies'; 'Tramwei for going to market, 1 anna 6 pies'; 'Grain to sparrow (Canary seed) 1 anna 3 pies'; 'Making white of master's hat, 5 pies'. And when at last you find a charge big enough to lay hold of, the imperturbable man proceeds to explain how, in the case of that particular item, he was able, by the exercise of a little forethought, to save you 2 annas and 3 pies. I have struggled against these accounts and know them. It is vain to be indignant. You must just pay the bill, and if you do not want another, you must make up your mind to be your own treasurer. You will fall in your boy's estimation, but it does not follow that he will leave your service. The notion that every native servant makes a principle of saving the whole of his wages and remitting them monthly to Goa or Nowsaree, is one of the ancient myths of Anglo-India. The ordinary boy, I believe, is not a prey to ambition and, if he can find service to his mind, easily reconciles himself to living on his wages, or, as he terms it in the practical spirit of oriental imagery, 'eating' them. The conditions he values seem to be: permanence, respectful treatment, immunity from kicks and cuffs and from abuse, especially in his own tongue, and, above all, a quiet life, without *kitkit*, which may be vulgarly translated 'nagging'.

There is one other thing on which he sets his childish heart. He likes service with a master who is in some sort a Burra Saheb. He is by nature a hero worshipper—and master is his natural hero. The saying, that no man is a hero to his valet, has no application here. In India, if you are not a hero to your own boy, I should say, without wishing to be unpleasant, that the probabilities are against your being a hero

---

Pie(s)   smallest unit of old Indian currency: 3 pies = 1 paisa, 4 paise = 1 anna, 16 annas = 1 rupee; a pie was worth less than half a farthing

to anybody. It is very difficult for us, with our notions, to enter into the boy's beautiful idea of the relationship which subsists between him and master. To get at it at all we must realize that no shade of radicalism has ever crossed his social theory. 'Liberty, Equality, and Fraternity' is a monstrous conception, to which he would not open his mind if he could. He sees that the world contains masters and servants, and doubts not that the former were provided for the accommodation of the latter. His fate having made him a servant, his master is the foundation on which he stands. Everything, therefore, which relates to the well-being, and especially to the reputation, of his master, is a personal concern of his own. Living always under the influence of this spirit, the boy never loses an opportunity of enforcing your importance, and his own as your representative. When you are staying with friends, he gives the butler notice of your tastes. If tea is made for breakfast, he demands coffee or cocoa; if jam is opened, he will try to insist on marmalade. At a hotel he orders special dishes. When your wife is away, he seems to feel a special responsibility, and my friend's boy, when warning his master against an unwholesome luxury, would enforce his words with the gentle admonition—'Missis never allowing, sir'.

It is this way of regarding himself and his master which makes the boy generally such a faithful servant. Veracity is the point on which he is weakest, but even in this there are exceptions. My last boy was curiously scrupulous about the truth, and would rarely tell a lie, even to shield himself from blame, though he would do so to get the *hamal* into a scrap. I regret to say that the boy has flaws. His memory is a miracle; but just once in a way, when you are dining at the club, he lays out your clothes nicely without a collar. He sends you off on an excursion to Matheran, and packs your box in his neat way; but instead of putting in one complete sleeping suit, he puts the upper parts of two, without the nether and more necessary portions. It is irritating to discover, when you are dressing in a hurry, that he has put your studs into the upper flap of your shirt front; but I am not sure it does not try your patience more to find out, as you brush your teeth, that he has replenished your tooth-powder box from a bottle of Gregory's mixture. But dhobie day is his opportunity. He first delivers

---

Hamal male servant, equivalent to British maidservant

the soiled clothes by tale, diving into each pocket to see if you have left rupees in it; but he sends a set of studs to be washed. Then he sits down to execute repairs. He has an assorted packet of metal and cotton buttons beside him, from which he takes at random. He finishes with your socks, which he skilfully darns with white thread, and contemplates the piebald effect with much satisfaction; after which he puts them up into little balls, each containing a pair of different colours. Finally he will arrange all the clean clothes in the drawer on a principle of his own, the effect of which will find its final development in your temper when you go in haste for a handkerchief. I suspect there is often an explanation of these things which we do not think of. The poor boy has other things on his mind besides your clothes. He has a wife, or two, and children, and they are not with him. His child sickens and dies, or his wife runs away with someone else, and carries off all the jewelry in which he has invested his savings; but he goes about his work in silence and we only remark that he has been unusually stupid the last few days.

So much for the boy in general. As for your own particular boy, he must be a very exceptional specimen if he has not persuaded you long since that, although boys in general are a rascally lot, you have been singularly fortunate in yours.

Ibid.

The Paragon

One story more to teach us to judge charitably. A lady was inveighing to a friend against the whole race of Indian cooks as dirty, disorderly, and dishonest. She had managed to secure the services of a Chinese cook, and was much pleased with the contrast. Her friend did not altogether agree with her, and was sceptical about the immaculate Chinaman. 'Put it to the test,' said the first lady; 'just let us pay a visit to your kitchen, and then come and see mine.' So they went together. What need to describe the Bobberjee-Khana? They glanced round, and hurried out, for it was too horrible to be endured long. When they went to the Chinaman's kitchen, the contrast was indeed striking. The pots and pans shone like silver, the table was positively sweet; everything was in its proper place, and Chang himself, sitting on his box, was washing his feet in the soup tureen!

Ibid.

A Ballad of the Bawarchikhana

The *bandobast* has turned to dust, departed are the staff
who ruled our lives and spoilt our wives, but often made us laugh—
a fitting epitaph!

Their presence felt where'er we dwelt: the discreet coughs and sighs,
the low salaams, the itching palms, the shrewd observant eyes,
the loyalty and lies

were all a part, close to the heart, of India's strange delight,
which we regret, while we forget the heat, the dust and bite
of insects in the night.

Old servants true pass in review as we relive those years,
but, best of all, I still recall a prince of profiteers,
a rogue who had no peers;

a man of guile with ready smile, an artist and a crook—
the kind you'd find all over *Hind*, described in many a book—
a not uncommon cook.

Abdullah Khan, our *Khansaman*, obeyed the Prophet's charge
to kneel and pray an hour a day, which left him ample marge
to prey upon the *Raj*.

Who could forget who once had met that mirror of his race—
the grave repose, the spotless clothes, the slow unhurried pace,
the strongly bearded face—

the look of grief if called a thief, the air of pained surprise
if someone noted he'd promoted all the humble *pies*
to *paise* in disguise?

A *naukar* full of *dhoka* and exceedingly *chalak*,
whose measured *seer* was nowhere near but well below the mark
by more than a *chittack*.

The kitchen scales could tell some tales, but let us not forget
how Abdul, roused from deepest drowse, so nonchalantly met
the unexpected threat—

Bawarchikhana kitchen   *bandobast* household management   *pies* smallest
unit in the old currency; see p. 112   *naukar* servant   *dhoka* deceit   *chalak* sly,
smart, untrustorthy   *seer* 2 lbs   *chittack* 2 oz

'Dinner for ten instead of four in half an hour? *Ji-Huzoor*.'
Whate'er the means behind the scenes, we knew the end was sure—
dinner for ten or more.

Or how in camp, by feeble lamplight in a smoky tent,
he'd improvise some rare surprise and recipes invent
astonishingly blent.

On days of sport, if guns were short, he followed in the truck—
a man inspired, he never tired of hoping for the luck
to shoot a sitting duck.

His *chits* explained he had been trained by famous *burra-mem*,
and though acquired, so it transpired, by devious stratagem,
he dearly treasured them.

So, well equipped with spurious script, he rose from humble 'Sub'
to dizzy heights where lesser lights prepared the actual grub—
as happened at our Club.

He rose at dawn and on the lawn performed his morning prayers;
and combed his beard and sometimes smeared fresh *henna* on
    the hairs
to falsify the years.

At six precise a lowly *syce* wheeled out an ancient 'bike',
which A. bestrode in cautious mode, his chin projecting like
a partriarchal pike.

Pyjamas stowed in socks he rode to shop in the bazaar,
a boy at heel beside his wheel, or propped on handlebar—
(for *chokra* annas *char*).

When he got back at eight, alack!—our day had just begun,
while Abdul smoked his hookah, joked and sheltered from the sun,
his daily labours done.

Reclining on his bed at ease, instead of counting sheep,
he totalled chits and perquisites with concentration deep
until he fell asleep.

Ji-huzoor Yes, Master   *chit* reference, note of hand, memo of account   *burra-mem* senior (English) lady   *henna* a red vegetable dye   *syce* (*salis*) groom   *chokra* small boy   *char* four

And while he snored, his minions pawed and cooked the food
    he'd bought
by rule of thumb, cooked to a crumb, without a second thought
to lessons he had taught.

The *masalchee* who reeked of *ghee*, the *bhisti* with his stoop,
the sweeper's wife who all her life had suffered from the croup,
all helped to stir the soup.

The hairs that stole into the bowl were not from Abdul's head—
they were too long, the colour wrong for Abdul to have shed—
for Abdul's hair was red.

The egg, long nursed, which came in first, the fowl which also ran,
the tea-leaves stewed instead of brewed, to give a deeper tan,
the porridge made of bran,

the half-washed dish, the muddy fish, the goat as tough as leather,
the basket pudding which collapsed, if cloudy was the weather
for several hours together:

All these and more, perforce we bore; we queried every bill,
and swore amain, but all in vain—the net result was nil—
Abdullah knew his drill.

Quite unconcerned, he neatly turned the sharp edge of our wit,
and parried point with counter-point and documented chit,
until he was acquit.

We could have fired Abdullah, hired another in his place—
who might be worse—that was the curse we always had to face;
and so to meet the case,

He got the sack, but soon came back to end as he began;
(some day we'd change the kitchen range—the system—not the man)
and so the legend ran:

'A first class cook and, as a crook, a bigger crook than most;
to any friend we recommend Abdullah for the post.'
Farewell familiar ghost!

Ghee clarified butter   basket pudding fruit trifle inside a burnt sugar casing

The Raj has handed over charge and Abdul's gone as well:
I wonder who warms up the stew and burns the toast in Hell?
Abdullah would excel!

It may be A . . . I cannot say, but I prefer to guess
he's found some creep in which to sleep, a man of his address,
behind the Devil's Mess.

R.V. Vernède, ICS

The Mussaul (Masalchi) or Man of Lamps

The Mussaul's name is Mukkun, which means butter, and of this commodity I believe he absorbs as much as he can honestly or dishonestly come by. How else does the surface of him acquire that glossy oleaginous appearance, as if he would take fire easily and burn well? I wish we could do without him! The centre of his influence, a small room in the suburbs of the dining-room, which he calls the dispence, or dispence-khana (bottle-khana, pantry), is a place of unwholesome sights and noisome odours, which it is good not to visit unless as Hercules visited the stables of Augeas. The instruments of his profession are there, a large *handie* (wide earthenware basin), full of very greasy water, with bits of lemon peel and fragments of broken victuals swimming in it, and a short stout stick, with a little bunch of foul rag tied to one end of it. Here the mussaul sits on the ice *numda* while we have our meals, and as each plate returns from the table, he takes charge of it, and transfers to his mouth whatever he finds on it, for he is of the omnivora, like the crow. Then he seizes his weapon of offence, and, dipping the rag end into the handie, gives the plate a masterly wipe, and lays it on the table upside down, or dries it with a damask table napkin. . . . When the mussaul has disposed of the breakfast things in this summary way, he betakes himself to the great work of the day, the polishing of the knives. He first plunges the ivory handles into boiling water, and leaves them to steep for a time, then he seats himself on the ice again, and arranging a plank of wood in a sloping position, holds it fast with all the energy which he has saved by the neglect of other duties. Hour after hour the squeaky, squeaky, squeaky sound of the board plays upon your nerves, not the nerves

---

Numda coarse woollen rug

of the ear, but the nerves of the mind, for there is more in it than the ear can convey. Every sight and every sound in this world comes to us inextricably woven into the warp which the mind supplies, and, as you listen to that baleful sound, you seem to feel with your finger points the back of each good new knife getting sharper and sharper, and to watch its progress as it wears away at the point of greatest pressure, until the end of the blade is connected to the rest by a narrow neck; which eventually breaks, and the point falls off, leaving the knife in that condition so familiar to us all, when the blade, about three inches long, ends in a jagged square point, the handle, meanwhile, having acquired a rich orange hue. Oh, those knives! those knives!

Etymologically Mukkun is a man of lamps and, when he has brushed your boots and stowed them away under your bed, putting the left boot on the right side and vice versa, in order that the toes may point outwards, as he considers they should, then he addresses himself to this part of his duty. Old Bombayites can remember the days of cocoanut, when he had to begin his operations during the cold season by putting a row of bottles out in the sun to melt the frozen oil; but kerosine has changed all that, and he has nothing to do but to trim the wick into that fork-tailed pattern in which he delights and which secures the minimum of light with the maximum destruction of chimneys, to smear the outside of each lamp with his greasy fingers, to conjure away a gallon or so of oil and to meet remonstrance with a child-like query, 'Do I drink kerosine oil?'. Then he unbends, and gives himself up to a gentle form of recreation in which he finds much enjoyment. This is to perch on a low wall or big stone at the garden gate, and watch the carriages and horses as they pass by. Other mussauls, ghora-wallas, and passing ice coolies stop and perch beside him, and sometimes an ayah or two, with a perambulator and its weary little occupant, grace the gathering. I suppose the topics of the day are discussed, the chances of a Russian invasion, the dearness of rice, and the events which led to the dismissal of Mr Smith's old mussaul, Canjee. Then the time for the lighting of lamps arrives and Mukkun returns to his duties. Mukkun lives in dread of the devil. Nothing will induce him to pass at night by places where the foul fiend is known to walk, nor will he sleep alone without a light.

E.H. Aitken, *Behind the Bungalow*, 1889

That Dhobie

I have studied the dhobie and find him to be nothing else than an example of the abnormal development, under favourable conditions, of a disposition which is not only common to humanity, but pervades the whole animal kingdom. Destruction is so much easier than construction and so much more rapid and abundant in its physical results, that the devastator feels a jubilant joy in his work. The dhobie, dashing your cambric and fine linen against the stones, shattering a button, fraying a hem or rending a seam at every stroke, feels a triumphant contempt for the miserable creature whose plodding needle and thread put the garment together. This feeling is the germ from which the dhobie has grown. Day after day he has stood before that great black stone and wreaked his rage upon shirt and trouser and coat and coat and trouser and shirt. Then he has wrung them as if he were wringing the necks of poultry, and fixed them on his drying line with thorns and spikes, and finally he has taken the battered garments to his torture chamber and ploughed them with his iron, longwise and crosswise and slantwise, and dropped glowing cinders on their tenderest places. Son has followed father through countless generations in cultivating this passion for destruction, until it has become the monstrous growth which we see and shudder at in the dhobie. He is not tolerable. Submit to him we must, since resistance is futile; but his craven spirit makes submission difficult and resignation impossible. If he had the soul of a conqueror, if he wasted you like Attila, if he flung his iron into the clothes-basket, and cried *Voe victis*, then a feeling of respect would soften the bitterness of the conquered; but he conceals his ravages like the white ant, and you are betrayed in your hour of need. When he comes in, limping and groaning under his stupendous bundle and lays out *khamees, patloon* and *pajama*, all so fair and decently folded and delivers them by tale in a voice whose monotonous cadence seems to tell of some undercurrent of perennial sorrow in his life, who could guess what horrors his perfidious heart is privy to? Next morning, when you spring from your tub and shake out the great jail towel which is to wrap your shivering person in its warm folds, lo! it yawns from end to end. There is nothing but a border, a fringe, left. You fling on your clothes in unusual haste, for

---

Khamees (kameez) shirt   *patloon* trousers   *pajama* pyjamas

it is mail day morning. The most indispensable of them all has scarcely a remnant of a button remaining. You snatch up another which seems in better condition and scramble into it; but in the course of the day a cold current of wind, penetrating where it ought not, makes you aware of what your friends behind your back have noticed for some time, viz. that the starch with which a gaping rent has been carefully gummed together, that you might not see it, has melted and given way. The thought of these things makes a man feel like Vesuvius on the eve of an eruption; but you must wait for relief until dhobie day next week, and then the poltroon has stayed at home, and sent his brother to report that he is suffering from a severe stomach-ache. When the miscreant makes his next appearance in person, he stands on one leg, with joined palms and a piteous bleat, and pleads an alibi. He was absent about the marriage of a relation, and his brother washed the clothes. So your lava falls back into its crater, or, I am afraid, more often overflows the surrounding country. My theory of the dhobie is a mere speculation—I do not pretend to have established it by scientific observation, and am very tolerant towards other theories, especially one which is supported by many competent authorities, and explains the dhobie by supposing a league between him, the dirzee and the boy. I think a close investigation into the natural history of the shirt would go far to establish this theory as at least partially true. In spite of the spread of 'Europe shops', the shirt is still abundantly produced from the vernacular dirzee sitting cross-legged in the verandah and each shirt will be found to furnish him, on the average, with half a week's lucrative employment. From his hand it passes to the dhobie and returns with the buttons wanting, the button holes widened to great gaping fish-mouths and the hems of the cuffs slightly frayed. The last is the most significant fact, because we discover that the hem has been made with the least possible margin of cloth, as if to facilitate the process of fraying. As we know that economy of material is not an object with the dirzee, it has been maintained that there is some connection here. Next the shirt passes into the hands of the boy, who takes his scissors and carefully pares the ragged edges of the cuffs and collar. A few rotations of dhobie and boy reduce the cuffs to the breadth of an inch, while the collar becomes a circular saw which threatens to take your head off. Then you fling the shirt to your boy, and the dirzee is in requisition again. Observation of white trousers will lead to similar results. Between dhobie's fury and boy's repairs, the ends of the legs retreat steadily upwards to your knees, and by the time the boy inherits them, they are just his length. Remember, I do not say I believe in this explanation of the dhobie. I give it for what it is worth. The subject is interesting and practical.

Did you ever open your handkerchief with the suspicion that you had got a duster in your pocket by mistake, till the name of De Souza blazoned on the corner showed you that you were wearing someone else's property? An accident of this kind reveals a beneficent branch of the dhobie's business, one in which he comes to the relief of needy respectability. Suppose yourself, if you can, to be Mr Lobo enjoying the position of first violinist in a string band which performs at Parsee weddings and on other festive occasions. Noblesse oblige; you cannot evade the necessity for clean shirt-fronts, ill able as your precarious income may be to meet it. In the circumstances a dhobie with good connections is what you require. He finds you in shirts of the best quality at so much an evening. . . . You need to keep no clothes except a greenish surtout and pants and an effective necktie. In this way the wealth of the rich helps the want of the poor without their feeling it or knowing it. Sometimes, unfortunately, Mr Lobo has a few clothes of his own and then the dhobie may exchange them by mistake. But, if you occasionally suffer in this way, you gain by another, for Mr Lobo's family are skilful with the needle and I have sent a torn garment to the washing, which returned skilfully repaired. I suspect I am getting bitter and ironical. . . . It is quite possible that, in the mild twilight of life, in the old country, I shall find myself speaking benevolently of the dhobie and secretly wishing I could hear his plaintive monotone again counting out my linen at four rupees a hundred.

Ibid.

The Bheestee

The malee has an ally called the bheestee. Bihisht in the Persian tongue means Paradise, and a bihishtee is therefore an inhabitant of Paradise, a cherub, a seraph, an angel of mercy. He has no wings; the painters have misconceived him; but his back is bowed down with the burden of a great goat-skin swollen to bursting with the elixir of life. He walks the land when the heaven above him is brass and the earth iron, when the trees and shrubs are languishing and the last blade of grass has given up the struggle for life, when the very roses smell only of dust, and all day long the roaring dust devils waltz about the fields, whirling leaf and grass and corn stalk round and round and up and away into the regions of the sky; and he unties a leather thong which chokes the throat of his goat-skin just where the head of the poor old goat was cut off, and straight-away, with a life-reviving gurgle, the steam called *thunda panee*, gushes forth, and plant and shrub lift up their heads and the garden smiles again. The dust also on the roads is laid and a grateful incense rises from the ground, the sides of the water chatty grow dark and moist and cool themselves in the hot air, and through the dripping interstices of the *khas-khas* tattie a chilly fragrance creeps into the room. . . . Nay, the seraph finds his way to your very bathroom, and discharging a cataract into the great tub, leaves it heaving like the ocean after a storm. When you follow him there, you will thank that nameless poet who gave our humble Aquarius the title he bears. Surely in the world there can be no luxury like an Indian 'tub' after a long march, or a morning's shooting, in the month of May. I know of none. . . . There is first the enfranchisement of your steaming limbs from gaiter and shooting boot, buckskin and flannel; then the steeping of your sodden head in the pellucid depth, with bubaline snortings and expirations of satisfaction; then, as the first cold stream from the 'tinpot' courses down your spine, what electric thrills start from a dozen ganglia and flush your whole nervous system with new life! Finally there is the plunge and the wallow and the splash, with a feeling of kinship to the porpoise in its joy, under the influence of which the most silent man becomes vocal and makes the walls of the narrow *ghoosul-khana* resound with amorous or patriotic song. A flavour of sadness mingles here, for you must come out at last, but the ample gaol towel receives you in its warm embrace and a glow of contentment pervades your frame, which seems like a special preparation for the soothing touch of cool, clean linen, and white duck, or smooth khakee. And even before the voice of the butler is heard at the door, your olfactory nerves, quickened by the tonic of the tub, have told you what he is going to say.

Some people in India always bathe in hot water, not for their sins, but because they like it. At least so they say, and it may be true, for

---

Thunda panee cold water   *khas-khas* window screens filled with the aromatic roots of *khas-khas* grass, used to cool rooms in summer   *ghoosul-khana* bathroom

I have been told that you may get a taste even for drinking hot water if you keep at it long enough.

The bheestee is the only one of all our servants who never asks for a rise in pay on account of the increase of his family. But he is not like the other servants. We do not think of him as one of the household. We do not know his name, and seldom or never speak to him; but I follow him about, as you would some little animal, and observe his ways. I find that he always stands on his left leg, which is like an iron gate-post, and props himself with his right. I cannot discover whether he straightens out when he goes home at night, but when visible in the daytime, he is always bowed, either under the weight of his *mussuk*, or the recollection of it. The constant application of that great cold poultice must surely bring on chronic lumbago, but he does not complain. I notice, however, that his waist is always bound about with many folds of unbleached cotton cloth and other protective gear. The place to study him to advantage is the *bowrie*, or station well, in a little hollow at the foot of a hill. Of course there are many wells, but some have a bad reputation for guinea-worm, and some are brackish, some are jealously guarded by the brahmins, who curse the bheestee if he approaches, and some are for low caste people. This well is used by the station generally, and the water of it is very 'sweet'. Any native in the place will tell you that if you drink of this well you will always have an appetite for your meals and digest your food. It is circular and surrounded by a strong parapet wall, over which, if you peep cautiously into the dark abyss, you may catch a sight of the wary tortoise, which shares with a score or so of gigantic frogs the task of keeping the water sweet. It was introduced for the purpose by a thoughtful bheestee; the frogs fell in. Wild pigeons have their nests in holes in the sides of the well. Here, morning and evening, you will find the bheestees of the station congregated, some coming and some going, but most standing on the wall and letting down their leather buckets into the water. As they begin to haul these up again hand over hand, you will look to see them all topple head foremost into the well, but they do not as a rule. It makes an imaginative European giddy to look down into that Tartarean depth; but then the bheestee is not imaginative. At the same time, the demand for water increases, for man is thirsty and the ground parched. So the toils of the poor

---

Mussuk goat-skin bag filled with water from a well

bheestee march *pari passu* with the tyranny of the climate, and he grows thin and very black. Then, with the rain, his vacation begins. Happy man if his master does not cut his pay on the ground that he has little to do. We masters sometimes do that sort of thing.

I believe the mussuk-bearer is the true and original bheestee, but in many places, as wealth and luxury have spread, he has emancipated his own back and laid his burden on the patient bullock, which walks sagaciously before him, and stops at the word of command beside each flower-pot or bush. He treats his slave kindly, hanging little bells and cowries about its neck. If it is refractory he does not beat it, but gently reviles its female ancestors. I like the bheestee and respect him. As a man he is temperate and contented eating *bajree* bread and slaking his thirst with his own element. The author of Hobson-Johnson says he never saw a drunken bheestee. And, as a servant, he is laborious and faithful, rarely shirking his work, seeking it out rather. For example, we have a bottle-shaped filter of porous stoneware, standing in a bucket of water, which it was his duty to fill daily; but the good man, not content with doing his bare duty, took the plug out of the filter and filled it too! And all the station knows how assiduously he fills the rain gauge. But what I like best in him is love of nature. He keeps a tame lark in a very small cage, covered with a dark cloth that it may sing, and early in the morning you will find him in the fields, catching grasshoppers for his little pet. I am speaking of a Mahommedan bheestee. You must not expect love of nature in a Hindoo.

Ibid.

Hurree the Dirzee

A warm altercation is going on in the verandah. A little human animal, with a very large red turban on his little head, stuck full of pins and threaded needles, stands on all fours over a garment of an unmentionable kind, which I recognize as belonging to me, and a piece of cloth lies before him, out of which he has cut a figure resembling the said garment. The scissors with which the operation was performed are still lying open upon the ground before him. His head is thrown so far back that the great turban rests between his shoulder-blades, his

---

Bajree bred bread made of *bajra* (kind of millet) flour

Hurree the Dirzee

brow is corrugated with perplexity, his mouth a little open, as if his lower jaw could not quite follow the rest of his upturned face. Hurree cannot know much about toothache. What would I not give for that set of incisors, regular as the teeth of a saw, and all as red as a fresh brick! I suppose the current quid of *pan supare* is temporarily stowed away under that swelling in the left cheek, where the fierce black patch of whisker grows. The survival of a partial cheek pouch in some branches of the human race is a point that escaped Darwin. But I am digressing into reflections. To return: a lady is standing over the quadruped and evidently expressing serious displeasure in some form of that domestic language which we call Hindoostanee, with variations. The charge she lays against him seems to be that he has, in disregard of explicit instructions and defiance of common sense, made a blunder to which her whole past experience of India furnishes no parallel, and which has resulted in the total destruction of a whole piece of costly material, and the wreck of a garment for want of which the saheb (that is myself) will be put to a degree of inconvenience which cannot be estimated in rupees, and will most certainly be provoked to an outbreak of indignation too terrible to be described. So little do we know ourselves! I had no idea I harboured such a temper. However, Hurree does not tremble, but pleads that it was necessary to make the garment 'leetle silope', and though he admits that the slope is too great, he thinks the mistake can be remedied, and is pulling the cloth to see if it will not stretch to the required shape. Falling this, he has other remedies of a technical kind to suggest. I do not understand these matters, and cannot interpret his argument, but he puts his fingers on the floor and flings himself lightly to the other side of the cloth, to point out where he proposes to have a 'false hame', or some other device. She rejects the proposal with scorn, and again impresses him with the consequences of his wicked blunder. At last I am glad to see that a compromise is effected and the little man settles himself in the middle of a small carpet and locks his legs together so that his shins form an X and he sits on his feet. In this position he will ply the needle for the rest of the day at a rate inversely proportional to the distance of his mistress. When she retires for her afternoon siesta the needle will nap too. Then he will take out a little Vade Mecum,

---

Pan suparee combination of betel leaf and areca nut, for chewing   'leetle silope' 'little slope', which means cut at a slight angle   'false hame' false hem

which is never absent from his waist-band, and unroll it. It is many coloured and contains little pockets, one for fragments of the spicy areca, one for a small tin box which contains fresh lime, one for cloves, one for cardamoms, and so on. He will put a little of this and a little of that into his palm, then roll them all up in a betel leaf out of another pocket, and push the parcel into his mouth. Thus refreshed he will go to work again, not, however, upon the garment to which he is now devoted, but upon a roll of coloured stuffs on which he is at the present moment sitting. You see, times are hard and Hurree has a large family, so he is obliged to eke out his salary by contract work for the mussaul. His work suffers from other interruptions. When the carriage of a visitor is heard, he has to awaken the chuprassie on duty at the door, and on his own account he goes out to drink water at least as often as the chuprassie himself. As the day draws near its close, he watches the shadow like a hireling, and when it touches the foot of the long armchair, he springs to his feet, rolls up his rags and threads into a bundle and trips gaily out. As he does so you will observe that his legs are bandy. This is the result of the position in which he spends his days.

This is how we clothe ourselves in our Indian Empire. Our smooth and comfortable khakee suits, our ample pyjamas, the cool white jackets in which we dine, in this way are they brought about. But you must not allow yourself to think of the dirzee simply as an agency for producing clothes. Life is not made up of such simplicities. Let imagination blot out the dirzee. Remove him from the verandah. Take up his carpet and sweep away the litter. What a strange void there is in the place! Eliminate him from a Lady's day. Let nine o'clock strike, but bring no stealthy footstep to the door, no muffled voice making respectful application for his *kam*. From nine to ten breakfast will fill the breach, and you may allow another hour for the butler's account and the *godown*; but there is still a yawning chasm of at least two hours between eleven and *tiffin*. I cannot bridge it. Imagination strikes work. . . . In the spirit of fair play, however, I must mention that my wife does not endorse all this. On the contrary, she tells me (she has a terse way of speaking) that it is 'rank bosh'. She declares that the dirzee is the bane of her life, that he is worse than a fly, that she cannot sit down to the piano for five minutes but he comes buzzing round for black thread, or white thread, or mother-of-pearl buttons,

---

Kam work, job   *godown* store shed   *tiffin* lunch

or hooks and eyes, that every evening for the last month he has watched her getting ready for a drive, and just as her foot was on the carriage step, has reminded her with a cough, that his work was finished and he had nothing to do. If she could only do without him, she would send him about his business and be the happiest woman in the world, for she could devote the whole day to music and painting and the improvement of her mind. Of course I assent. That is a very commendable way of thinking about the matter. But, as an amateur philosopher, I warn you never to let yourself get under practical bondage to such notions. I tell you when you betake yourself to music or painting, carpentry or gardening, as a means of getting through the day, you are sapping your mental constitution and shortening your life: unless you are sustained by more than ordinary littleness of mind you will never see threescore and ten.

Maria Graham tells us that in her time the 'Dirdjees', or tailors, in Bombay were 'Hindoos' of respectable caste, but in these days the Goanese, who has not capacity to be a butler or cook, becomes a dirzee, and in Bombay I have seen *bunniah* dirzees. Hurree can hold his own against these, I doubt not, but the advancing tide of civilization is surely crumbling down his foundations. It is not only the 'Europe' shop in Bombay that takes the bread out of his mouth, but in the smallest and most remote stations, Narayan, 'Tailor, Outfitter, Milliner and Dressmaker' hangs out his sign-board, and under it, pale consumptive youths bend over their work by lamplight and sing the song of the shirt to the whirr-rr-rr of sewing machines. And as Hurree goes by, his prophetic soul tells him that his son will not live the happy and independent life which has fallen to his lot. But he has a bulwark still in the dhobie, for the 'Tailor and Outfitter' will not repair frayed cuffs, and the sewing machine cannot put on buttons. And Hurree is not ungrateful, for I observe that, when the dhobie delivers up your clothes in a state which requires the dirzee, the dirzee always gives them back in a condition which demands the dhobie.

Ibid.

Buggoo, the Chowkeydar

His life is a watch or a vision
between a sleep and a sleep
      Atalanta in Calydon

Bunniah (bania) Indian merchant   *chowkeydar* watchman

And who was Buggoo? Buggoo was a chowkeydar and Sudhoo's neighbour. That wigwam is Buggoo's house; his wants are very few. Besides, he seldom sees his house by daylight—the crank walls and the latticed roof look well enough by night; so Buggoo is contented with his house, and as he sallies forth to his work, he sings a hideous refrain at the pitch of his voice, answering cheerily the owls. The chowkeydar is an animal *sui generis*, and the only one species of his genus. The family has but a single order—chowkeydars—and besides them there is no other, neither any varieties. His childhood is a tradition. Perhaps in early youth he was a pea-boy, and so acquired a taste for grotesque shouting; but it is more reasonable to suppose that he never was a boy. He was born adult. He exists by night, and his days are divided into moonlight and pitchy darkness. For one half of his life he has no shadow. He knows of the sun but is not intimate with it; the constellations he is familiar with, taking his time from the rise and decline of the Hesperids or Orion. For the periods of his working hours Nature has provided him a chronometer. If when he comes to his work the bats are still fluttering in and out of the rafters, he says 'I am up betimes, 'tis early'; the Great Bear is aslope, and he says 'The day's work is half over', and when the jackal cries the third time at the break of dawn he says 'I feel sleepy, night approaches.' And in this he cuts himself off from his kind, sets himself apart from humanity, in that at early morning he goes yawning home like the beasts of the forest and not forth to his work like the sons of men. He knows what the sun is like—to see a festival he has sat up the live-long day. More than this Buggoo does not know. Bats are his sparrows and moths his flies: an afternoon is as secret from him as the Feast of Fatima or the system of computation by quaternions, and of the sun at midday he speaks as we speak of the Southern Cross. He holds it his duty to sleep all day, because he has been up all night, though he sleeps all night to shirk his duty. Now and again he wakes up and clears his throat to let the world know it, or yells in answer to some distant friend; but he does little more. When he comes first he seizes his iron-shod staff of office, and striking it as he goes against the dull ground makes the circuit of the house. Beneath the porch he loiters with the servant who is sitting up to see his master home—saunters round the corner, and as he passes each bedroom door, startles the night with an unearthly cry, putting the jackals to shame, or breaks off suddenly in the middle to choke in a reassuring manner. He then coughs defiantly, hiccups, and passes on—tramp! tramp! 'And Beauty sate in the hall waiting for him, and at last she heard him coming, tramp! tramp! striking his club upon the ground, and suddenly round the corner came—the Beast'. The chowkeydar meanwhile has reached his blanket stretched out in a sheltered place, has scared away the cat which had taken possession of it, and is asleep.

Phil Robinson, In my Indian Garden, 1888

The Khidmatgar

Thou art so near and yet so far—
this line reminds me of his presence,
the silent white-robed *khidmatgar*,
whose costume scorns all iridescence.

Was ever such a noiseless tread?
Were ever hands so deft at waiting?
He needs not that a word be said,
your every wish anticipating.

He knows no English—so your friends
are wont to say, and yet I would not
swear that he never comprehends
some conversation that he should not.

A self-respect that's all his own
attends but to your eating, drinking;
regard his features—sculptured stone
without a trace of even winking.

How would he look (I once did muse)
if someone from behind should take him
with, say, a sharp pin thro' the 'trews'
or very violently shake him?

Unworthy thought! Let peace aye rest
upon those dignified calm features!
Throughout the world he is the best
of all domestic waiting creatures.

Of course he's got an inner life,
altho' I never seem to twig it,
behind the godowns where his wife
and children possibly pig it.

And he not always meek and mild,
may cut a very different figure
when crowing to an infant child
or ruling womankind with vigour.

Therefore, O white-robed khidmatgar
clean, silent, self-contained, quick-seeing,
I often wonder what you are—
inscrutable and shinx-like being?

    'Lunkah', *Whiffs Anglo-Indian*, 1891

The Chaprassi

At every house beside the door,
upon the smooth verandah floor
you see him sit or stand.
No belted Earl—his badge and belt
denote a service that has felt
full many a hard command,

and many a sharp rebuke; but which
by supple palms that daily itch
for balsam of rupees
is highly prized and him, who joins
those palms o'er six or seven coins
for monthly wage, one sees

supporting twenty of his tribe.
The vulgar thing we call a bribe
cannot account for this;
it shows he is the poor man's friend;
to him the banks will even lend,
so popular he is.

The world without whose thoughts afar
are hidden in the dense bazaar,
knows well how good and kind

he is, the servant of the State,
who helps its rulers small and great
to make up each his mind.

The judge upon the bench divines
from him, through some mysterious signs,
the truth which no-one saw;
the man who sues for half a lac
is safe with him behind his back
to chance the pleader's law.

And then petitions sent by post
are useless, as is known by most,
are seldom seen in fact;
But he can get them read, or bar
their being heard, and if they are
successful 'tis his tact.

Chaprassi, peon we call him, knave
too often, yet a useful slave,
whose tricks we never see;
to simple villagers a lord,
whose word unlocks the village hoard
of milk and flour and ghee.

The old traditions fade, but he
lives on, though scantier grows his fee—
Poor soul! let no-one flout him;
we're virtuous now and make a fuss
o'er little things, but which of us
can really do without him?

      Idem, *Whiffs Anglo-Indian*, 1891

No. XI The Red Chuprassie, or the Corrupt Lictor

The red chuprassie is our Colorado beetle, our potato disease, our
Home ruler, our cupboard skeleton, the little rift in our lute. The red-
coated chuprassie is a cancer in our Administration. To be rid of it
there is hardly any surgical operation we would not cheerfully
undergo. You might extract the Bishop of Bombay, amputate the

Governor of Madras, put a seton in the pay and allowances of the Lieutenant-Governor of Bengal, and we should smile.

The red chuprassie is ubiquitous; he is in the verandah of every official's house in India, from the Governor-General downwards; he is in the portico of every Court of Justice, every Treasury, every Public Office, every Government School, every Government Dispensary in the country. He walks behind the Collector; he follows the conservancy carts; he prowls about the candidate for employment; he hovers over the accused and the accuser; he haunts the Raja; he infests the tax-payer.

He wears the Imperial livery; he is to the entire population of India the exponent of British rule; he is the mother-in-law of liars, the high priest of extortioners, and the receiver-general of bribes.

Through this refracting medium the people of India see their rulers. The chuprassie paints his master in colours drawn from his own black heart. Every lie he tells, every insinuation he throws out, every demand he makes, is endorsed with his master's name. He is the archslanderer of our name in India.

He is not an individual—he is a member of a widely ramified society. There is no city in India, no mufussil-station, no little settlement of officials far up country, in which the chuprassie does not find sworn brothers and confederates. The cutcherry clerks and the police are with him everywhere; higher native officials are often on his side.

He sits at the receipt of custom in the Collector's verandah, and no native visitor dare approach who has not conciliated him with money. The candidate for employment, educated in our schools, and pregnant with words about purity, equality, justice, political economy and all the rest of it, addresses him with joined hands as 'Maharaj' and slips silver into his itching palm. The successful place-hunter pays him a feudal relief on receiving office or promotion, and benevolences flow in from all who have anything to hope or fear from those in power.

In the native states the chuprassie flourishes rampantly. He receives a regular salary through their representatives or vakils at the agencies, from all the native chiefs round about, and on all occasions of visits or return visits, durbars, religious festivals, or public ceremonial, he claims and receives preposterous fees. The Rajas, whose dignity is always exceedingly delicate, stand in great fear of the chuprassies.

They believe that on public occasions the chuprassies have sometimes the power of sicklying them over with the pale cast of neglect.

English officers who have become de-Europeanized from long residence among undomesticated natives, or by the habitual performance of petty ceremonial duties of an Oriental hue, employ chuprassies to aggrandize their importance. They always figure on a background of red chuprassies. Such officials are what Lord Lytton calls 'white baboos'.

Mr . . ., in his own artless way, once proposed legislation against chuprassies, I am told. His plan was to include them among the criminal classes and hand them over to . . . the Director-General of Thuggee and Dacoity; but this functionary, viewing the matter in a different light, made some demi-official representation to the Legal Member . . . and the subject was dropped.

A great Maharaja once told me that it was the tyranny of the Government chuprassies that made him take to drink. He spoke of them as the 'Pindarries of modern India'. He had a theory that the small pay we gave them accounted for heir evil courses. A chuprassie gets about eight pounds sterling a year. He added that if we saw a chuprassie on seven rupees a month living overtly at the rate of a thousand, we ought immediately to appoint him an attaché or put him in gaol.

I make a simple rule in my own establishment of dismissing a chuprassie as soon as he begins to wax fat. A native cannot become rich without waxing fat, because wealth is primarily enjoyed by the mild Gentoo as a means of procuring greasy food in large quantities. His secondary enjoyment is to sit upon it. He digs a hole in the ground for his rupees and broods over them, like a great obscene fowl. If you see a native sitting very hard on the same place day after day, you will find it worth your while to dig him up. Shares in this are better than the Madras gold mines.

In early Company days when the Empire was a baby, the European writers regarded with a kindly eye those profuse Orientals who went about bearing gifts; but Lord Clive closed this branch of the business, and it has been taken up by our scarlet runners or verandah parasites, in our name. Now, dear Vanity, you may call me a Russophile, or by any other marine term of endearment you like, if I don't think the old plan was the better of the two. We ourselves could conduct corruption decently; but to be responsible for corruption over which we exercise no control is to lose the credit of a good name and the profits of a bad one.

I hear that the Government of India proposes to form a mixed committee of Rajas and chuprassies to discuss the question as to whether native chiefs ever give bribes and native servants ever take them. It is expected that a report favourable to Indian morality will be the result. Of course Raja Joe Hookham will preside.

G.R. Alberigh-Mackay ('Sir Alibaba K.C.B.'),
Twenty-One Days in India, 1879

4
Sport

During the British era India produced sportsmen and athletes of international standard only in polo, hockey and squash. Cricket and tennis were to come later. There were no facilities for proper training, no bodies of experts to select promising youngsters. So this selection has only to do with the exploits of the British in India, and all on an amateur basis.

Riding

This, I consider, was the premier sport of India. The method of agriculture, the light soil, the absence of fences, fields being separated by a low mud wall with no physical demarcation between the lands of neighbouring landowners, all meant that the rider met with no obstruction, provided he kept out of standing crops. A crop of gram (lentil-like peas) would suffer no harm if one rode through it while it was still small. Also in the interest of his pony it was essential for a rider to keep out of *arhar* (a kind of 'pulse') fields after the crop had been cut as the stumps were solid and razor-sharp. Riding was the best way to learn the identity of all the crops. From the train one looked out on what appeared to be flat, almost featureless and dull plain. But riding over it revealed many features and contrasts which saved the ride from being dull. When touring, officers rode from one camp site to the next, so were able to deal on the spot with many complaints and petitions which would never have reached them at their headquarters.

Polo

The origins of this very ancient game are controversial. The most likely places are either China or Persia, possibly both. It must have been in existence before the fourth century BC.

About the eighth century AD it was introduced to Byzantium and probably from there to Languedoc in a version to be played on foot. The ball varied from straw and stones bound by rice paper in China to leather in Byzantium, bamboo root in Manipur and wood in Persia and elsewhere. For striking the ball two quite distinct instruments were devised: (1) a bat or racket, the business end being curved like a hockey stick with a lacrosse-type net but very much smaller, and (2) what by a paraphrase in English could be translated 'golf-stick', which developed later into the head of the polo-stick as we know it.

It seems that polo ceased to be the game of kings and princes in Persia and western Asia some time in the seventeenth century, but survived in a much rougher form in Manipur at the eastern end of the Himalaya and in Ladakh, Balti, Astor, Gilgit and Chitral at the western end of the Himalaya and in Tibet. It was from these areas and in particular from Balti and Manipur that polo was introduced to India, to Calcutta in 1862 and to the Punjab in 1864. In its mountain home it was a very rough game indeed. The British devised rules in the interests of safety, limited the height of ponies, regulated the width and thickness of goal posts and improved the surface of grounds. They also fixed the weight of balls and standardized the shape of sticks. Introduced from India to England and thence to the USA, the Argentine and elsewhere, polo has become almost exclusively a rich man's game. But in India a sensible arrangement was worked out. In addition to world-class players and tournaments, a system of low-handicap tournaments was introduced for those who could not afford a string of expensive polo ponies. Two sound and moderately fast ponies were the basic minimum for station polo and low-handicap tournaments. Some British Infantry teams competed in the low-handicap tournaments but for the most part the teams were made up on an *ad hoc* basis. But an Infantry Regiment, if it had a good enough team and ponies was not precluded from entering for the major tournaments. In 1894 the Durham Light Infantry won the Infantry Cup and went on to win 14 open tournaments, including the Inter-regimental three times, as well as the championships of 1898.

Players were rarely unseated, generally due to turning too sharply or because they were fouled (crossed too closely by an opponent when they themselves had the right of way). The incidence of fatal accidents to riders and ponies was very low.

Pigsticking

If polo was the sport of kings, pigsticking was the king of sports. The season was from November to June, give or take a month or two. Two sound and staunch ponies, three really sharp spears, two fellow-pigstickers to make up a heat (or better still, six riders to make up two heats) were the basic requirements. One had to be able to ride well, to be sharply observant, e.g. to be able to change direction at a split-second's notice to avoid a disused well and without losing track of the boar. The main skill and mental stimulation, only to be learnt by experience, were in mastering the habits of the wild pig and in particular its likely reactions and objectives when being chased. This last knowledge is also useful in fox-hunting but there the resemblance ends. In pigsticking the riders chase the boar themselves, not vicariously through hounds, and the animal chased is well armed and capable of putting up a good fight which he sometimes wins. The terrain over which the chase is conducted is rough and unknown to the riders. Like the fox, the wild pig do a lot of damage, in their case to the peasants' crops. Pigsticking is the best way of controlling their numbers because only mounted men can reach the scene of their depredations.

Some readers may have noticed a discrepancy between the advice given by Kipling for riding over rough and unknown country in 'The Peora Hunt', which was the chapter heading for his story called 'Cupid's Arrows' (*Plain Tales from the Hills*) and the stirring words of the anonymous 'Hunting Song'. The latter advocates the orthodox doctrine of pigsticking that whatever happens you must stick as close behind the pig as you can (a) so as not to lose him, and (b) to tire him. As regards the danger of a fall, the orthodox doctrine was that where a pig could go a horse and man could follow. The other doctrine, advocated in 'The Peora Hunt', was that whatever line the game you were hunting took, when this took you to the edge of a particularly dangerous-looking piece of cover, it was more sensible to make a detour and come out, hopefully, somewhere near where the pig would break from his cover. There was something to be said for both tactics, and the ideal in a heat was probably for one man to follow the orthodox line and for the other two to try to cut the boar off from the next cover of his choice. Because of the boar's formidable forward armament the incidence of minor and fatal injuries among horses was higher than in polo or hunting jackal or fox. But it was not high. Minor injuries to riders were not uncommon, but fatal accidents were very rare. The journal of the pigsticking clubs was *The Hoghunters Annual*, and the Blue Riband of pigsticking was to win the Kadir Cup in a competition held every year in sandy grasslands near the Ganges in Meerut district.

Hunting

This was run on the same lines as fox-hunting in England except that in India you chased a jackal. There were not more than a dozen packs, some very well known, e.g. The Quetta Hunt, The Delhi Hunt, the 'Ooty' Hunt, the Peshawar Vale Hunt and the Calcutta Hunt. Whereas women could only play polo with other women and were not allowed to go pigsticking, there were no such restrictions for hunting and many women rode to hounds. The constant anxiety was to keep the hounds fit and avoid rabies infection.

Golf

This in the plains only came into its own in the monsoon when the only other recreation was riding. The rules, with some variations to meet Indian conditions, were much the same as in England, but because of the terrain it was a very different game from golf in England. There were 'browns' instead of greens, it was impossible to construct proper fairways and even the best players hitting down the middle where the fairway ought to be might see their balls kick off at a tangent on landing. There was no way of keeping the course free of unwanted visitors, the odd cow or a small boy driving goats. It is known that a tiger walked onto one golf course to see for himself what was happening. Of golf courses in the hills I have very limited knowledge, but by hearsay they were to be compared with golf courses in England rather than with courses in the plains of India. One grew cunning on a plains course after a while, and this made it a less frustrating and more enjoyable game.

Shooting

India was, of course, famous for its shooting. In other countries the landowners took every possible step to preserve the game on their

Tiger Visits Golf Course

own estates. But, except for the native states, ninety per cent of the landowners in India were not at all interested in the game on their estates and a shooting party found it almost impossible to find out where one owner's land ended and another's began. On the ground, therefore, this meant they could shoot wherever they liked. Of the ten per cent who tried to preserve their game, some only did so in order to be able to invite the Viceroy or a Governor or other VIP to shoot over their land. But there were a few who wanted to shoot over their land for themselves. A predecessor in the British station who had discovered this, left notes for the civil and military officers so that a letter was sent to this landowner to say that no-one would shoot over his land without his permission. This often resulted in an invitation to a few civil and military officers to come as his guests to shoot over his land with him.

Small Game

For the cost of cartridges, and beaters if employed, you could have all the small game shooting you wanted—partridge, quail, black partridge, blue pigeon, green pigeon, sandgrouse, geese. Most districts had one or more *jheels* or lakes swarming with teal, duck and snipe. In his *Oriental Field Sports* the publisher/editor Edward Orme picked out snipe-shooting as being the most reprehensible, because the sportsman waded up to his waist in water under the midday sun. That together with copious draughts of claret or Madeira must sap his constitution and shorten his life. Though his description of the snipe's habitat may be true of Bengal, it is certainly not true of snipe-shooting up-country. There the favourite terrain for snipe was flooded fields where the water came no higher than just above one's ankles.

Big Game

Rhino and bison have long since disappeared from what was British India; a few may still be found in the Terai forests of Nepal. Contrary to some popular opinion, the British were not responsible for the disastrous drop in numbers of tigers which at one time threatened their survival. Under the Raj the Forest Department kept a very tight control. In the first enthusiasm of independence, controls were swept aside. At the same time inroads were made into tiger country by settlement schemes for landless peasants involving the felling of forest. The Indian Government has taken drastic steps and it now seems that the number of tigers has recovered again. In British times sportsmen also climbed up to considerable heights to shoot good specimens of the wild goats and sheep, *markhor*, ibex, *ovis ammon, bhurral* and *thar*, and also black and red bear. Because they were difficult to get at, I suspect that their numbers have kept up well. A census would be difficult to carry out.

Badminton

This was played in India under the 'Poona' rules (1876) long before lawn tennis, which did not reach India until the end of the nineteenth century.

Lawn Tennis

When it did come, it caught on at once and soon became the most popular game of all. It is not a game that lends itself easily to ridicule or satire and there is nothing anent tennis in this chapter.

Fishing

There was good fishing to be had in various parts of India. Trout had been acclimatized in Kashmir, Kulu and a few small lakes and rivers in the central Himalaya. The hardest fight and greatest thrill was provided by the *mahseer*, a large fish with a strong bone structure. It was fished for with a spoon.

Sailing

Where there was a suitable lake near a British station, sailing was popular and good sport.

Markhor a type of large, horned wild goat   *ovis ammon* great Tibettan wild sheep, with massive horns in the male   *bhurral* a link between sheep and goats, found at 12,000–14,000 ft   *thar* smaller wild goats, with smaller horns

Camping

It is quite legitimate to include camping under sport. Although camping was on duty, the environment, conditions, and opportunities made it possible to combine a little shooting with one's progress across the district. 'Lunkah', who from his verses under the title 'Camp' must have been in the ICS, rather gives the impression that the touring officer, on the day he moved camp, could spend the morning stalking black buck on foot or shooting duck and teal. In my province the unwritten rule was that an officer could take the days off during which there was a major Hindu, Muslim or Christian religious festival unless he had to stand by because of a threat of communal riot. An officer could also take the weekend off if work permitted. So in making out his cold weather tour programme the DC planned the route in such a way that at weekends he would be near a good shooting area. Occasionally the demands of work made it impossible to enjoy his weekend shooting. But the odds were that he could take a few hours off on a weekday to walk up snipe.

Looking back, one of the things I most enjoyed in India was the annual camping tour through the district. Every year the District Officer and his assistants had to tour through the district in such a way that the whole area of the district would be covered in two or at the outside three years. This touring was done in the cold weather between October and February, but more usually between October and Christmas. On one or two nights one might stay at some rest-house as a matter of convenience, but for most of the tour one camped in tents. The Government paid for enough transport—bullock carts and/ or camels—to carry two large tents (the shape and size of a medium-large marquee) and from six to eight small tents (*sholdaris*) for office, cookhouse and personal staff and furniture. The two large tents were truly magnificent, double-fly with the space under the outer fly covered in with folding panels (*kanauts*) to form an outer passage along the two long sides of the tent. The doorways were covered by *chicks* on the inside made of split bamboo and laced with twine. On cold nights or to keep out wild animals, the kanauts were drawn across the outside of the doorways. The inner lining was a straw-coloured cotton with a simple design in black lines printed on the cotton. The tent material and the inner design were, I understand, made by jail labour and were the same all over India. Cotton carpets (*daris*), again

Camping in the Moffusil

made in the jails, were laid over straw. These big tents had a smell all their own, which I found rather attractive. One tent was used as a dining/sitting room with a pantry at one end. The other tent was used as a bedroom-cum-bathroom. The picture is completed by a set of camp folding wooden furniture. One slept on a wooden and canvas bed, the canvas laced on one long and one short side to the wooden frame. One peculiarity of the doorway chicks was that you could see through them from inside but not from outside. Once I had discovered this I could have my bath with an easy mind, though there might have been several village children outside with their noses flattened against the outside of the chick. One camped almost always in a mango grove. These groves were planted all over northern India and probably elsewhere as well. The fruit was small and tasted of turpentine and was inedible. One would be tempted to think they were planted just to provide camping sites for civil and military officers, if one did not know that mango wood was highly prized by Indian carpenters and the foliage by small boys lopping fodder for their goats.

Some of the disputes dealt with in camp one knew about already, and they had been earmarked for disposal at various camp sites most convenient for the parties and one hoped, almost always in vain, inconvenient for their lawyers coming from *sadr* (capital of district). Bu there would be new disputes as well which one hoped to be able to settle on the spot. One learnt fairly early in one's service to distinguish between serious litigation and litigation indulged in as a game by peasants who had no other relief from their hard life except various religious festivals. This meant that a case was often simply a contest in one-upmanship and extravagant perjury.

The system invariably followed for moving camp was as follows: The night before one's own move, the dining tent and all the small tents except the *syce's* and the chief clerk's tents were struck after dinner and packed on bullock carts, and travelled through the night to the next site, where they were erected well before the *Sahib* arrived. At the old camp site, after an early cup of tea, the Sahib mounted one horse and one syce the other, the other syce having gone ahead to the next camp. If one was married, the syce left behind would have no mount unless one had three horses, and would follow with the other

---

Syce (sais) groom

tents later. The head clerk, himself a comfortable figure, followed on a still more well-fed pony. By keeping up an amble (*cuddam*) most of the way he was able to keep up, more or less, with the horse-riders and—the main reason for requiring his company—to be on the spot when any new petition was presented on the way. If one was not too much delayed on the way, the cook had 'brunch' ready on one's arrival; but if rather more heavily delayed by work on the way, by that marvellous bush telegraph which the western world has abandoned, the cook would know that it was *tiffin* that one would require. One might well get rather hot on the way between the two camp sites. In what seemed an incredibly short time the hot water for the bath was ready. A bonus was the delicious scent of the bath water heated over a log fire. And when one sat down to 'brunch' or tiffin, another bonus, if one had managed things properly, was the game-pie, renewed every day or so with the odd game bird shot for the pot.

In the course of one's cold weather camping one gained valuable information about several villages, village headmen and other officials such as the *patwari*s, and the *kachcha hal* or inside story of some current case. If one stopped to hear a petition or try a case there was a formal welcome before one could begin. One of the commonest forms which this welcome took was a glass of hot boiled milk and a hard-boiled egg, which the headman peeled with his own grubby fingers. I found hot boiled milk more refreshing than cold milk and, of course, it was safer to drink.

Before I leave the subject of sport, I want to refer again to *Oriental Field Sports*. Of the two authors, Capt. Williamson was the expert on game and hunting for game. He spent 20 years in Bengal. The other, Howitt, never went to India but was an artist who worked up the sketches sent to him by Williamson into watercolours which could then be reproduced as aquatints. I wonder if anyone else has noticed what I found in these splendid prints. All the tigers have rather curious heads. They seem to be sunk too much into their shoulders, i.e. no neck, and to be too long and heavy from ears to chin and too short from nose to neck. In nearly all the prints they have been given the most splendid ruffs. Perhaps this is what distinguishes the Royal Bengal Tiger from other tigers. I could hardly believe my eyes when

---

Tiffin lunch   *patwari* village accountant who kept and annually revised all land records in his circle which covered a number of villages.

I spotted that the elephants' trunks were depicted as split open on the underside from mouth to tip. This was only noticed clearly where the artist is standing directly in front of the elephant and the latter has raised its trunk in the air. Could it be that the gallant Captain's sketches were indeed so sketchy that his patient collaborator at home had to rely too much on his own imagination? Or—'Tell it not in Gath, publish it not in the streets of Askelon'—could it be possible that the Captain had never seen a dead tiger or a live elephant at close hand?

*

The three pieces are from Samuel Howitt and Thomas Williamson, *Oriental Field Sports* (pub. Edward Orme) 1805–7.

Plate XVII: A Tiger Springing on an Elephant

A curious circumstance occurred to a very worthy officer, Captain John Rotton, who was one of a numerous party, assembled for the purpose of tiger-hunting, and was mounted on a very fine male elephant, that, far from being timid, was very remarkable for a courage scarcely to be kept within the bounds of prudence. This singularly fine animal having after much beating of a thick grass, hit upon the tiger's situation, uttered his roar of vengeance, which roused the lurking animal, occasioning him to rise so as to be seen distinctly.

No sooner did the tiger show himself than Captain Rotton, with great readiness, bending his body a little to the left, took aim at him as he stood up, cross-wise almost close to the elephant's head. The elephant no sooner espied his enemy than he knelt down, as is common on such occasions, with the view to strike the tiger through with his tusks. At the same time the tiger, sensible of the device, as suddenly threw himself on his back; thereby evading the intended mischief and ready to claw the elephant's face with all four feet, which were thus turned upwards.

Now whether Captain Rotton had not been in the habit of joining in such rapid evolutions, or that the elephant forgot to warn him to hold fast, we know not, but it so happened that the delicate situation in which he was placed, while taking his aim, added to the quickness of the elephant's change of height forward, combined to project him, without the least obstruction, from his seat, landing him plump on the tiger's belly!

This was a species of warfare to which all parties were apparently strangers. The elephant, however fearless in other respects, was alarmed at the strange round mass, the Captain being remarkably fat, which had shot like a sack over his shoulders; while the tiger, judging it to be very ungentleman-like usage, lost no time in regaining his legs, trotting off at a round pace and abandoning the field to the victorious Captain (?).

Plate XVIII: Snipe-shooting

Of all the diversions which most certainly, and I may say most speedily sap the constitution, none can, in my mind, compare with snipe shooting. In India snipes lie best during the mid-day heats; and, for the most part, being found in broad quagmires, and abounding chiefly on the flat borders of jheels, or perhaps among the small islands in the interior, compel those who delight in this recreation to wade probably up to the waist in water; being alternately wet and dry, while a burning sun keeps the head and upper parts of the body in a state exactly the reverse of what the lower parts experience. The short time required to boil eggs suspended in a cloth and dipped repeatedly into boiling water, may serve to give some idea of the infallible result of such a combined attack on the principles of life. I could enumerate at least an hundred of my acquaintances who have sacrificed the most vigorous health to this destructive sport; but who, strange to say, never could shake off the fatal habitude of indulging in what they neither were, nor could be ignorant was destroying them by inches! Formerly it was not considered sufficient to indulge in this reputed diversion alone; custom had joined to it the equally baneful practice of drinking spirits in every mode of preparation.

With the exception of a few prudent men whose moderation rendered them contemptible in the opinion of the major part of us, who were greatly attached, not only to sport, but to every species of debauchery, I believe few quitted Berhampore in those days untainted by disease, or without some serious injury done to their constitutions. Happily an entire reform has long since taken place throughout India.

Snipe-Shooting in Bengal

Plate XXVI: Protection against the Heat of the Sun

The power of the sun is a great drawback to the pleasures of the field. Most sportsmen provide themselves with white turbans of quilted linen, which, covering the crown of their hats, keeps off the heat. The skin of a pelican, with the soft down adhering, like our swan skin powder-puffs, is, however, much lighter and cooler. Snipe-shooting is particularly insalubrious in India, being mostly in extensive swamps, and as the birds do not lay but in the middle of the day, the lower extremities are freezing, while the head is melting with heat. It is very unpleasant to follow game through quags, and to be sometimes nearly up to the neck in mud and water.

A facetious gentleman, Lieutenant George Boyd, who was an excellent and keen sportsman, whenever he went snipe-shooting, used to squat down in the first sufficient puddle he came to, so as to wet himself up to the neck; observing that he found it very unpleasant to be getting wet by inches, and that by this process he put himself out of pain. He did not live long!

The next three pieces are from *The Hoghunters Annual*.

Vol. I (1928): The Fighting Boar

God gave the man the horse to ride,
and steel wherewith to fight,
and wine to swell his soul with pride,
and women for delight;
but a greater gift than all these four
was when he made the fighting boar.

The horse is filled with a spirit rare,
his heart is bold and free;
the bright steel flashes in the air
and glitters hungrily.
But these were little use before
the Lord he made the fighting boar.

The ruby wine does banish care,
but it confounds the head,
the fickle fair is light as air

and makes the heart bleed red.
But wine nor Love can tempt us more
when we may hunt the fighting boar.

When Noah's big monsoon was laid,
the land began to rise again,
and then the first hog-spear was made
by the cunning hands of Tubal Cain.
The sons of Shem and many more
came out to hunt the fighting boar.

Those ancient Jew-boys rode like stinks,
they knew not rack nor fear,
old Noah knocked the first two jinks,
but, Nimrod got the spear.
And ever since those times of yore
true men have rode the fighting boar.

Drink then to women and to wine
though heart and head they steal,
but here's to steed and spear and swine,
a brimming glass, No Heel!
And humble thanks to God who saw
his way to make the fighting boar.

> (From the Log-book of the
> Royal Dragoons)

Vol. II (1929): Farewell to Simla

Farewell! farewell ambition,
I mock you now I'm free!
Farewell! dear Simla ladies,
you have no charms for me!

Farewell! black coats and uniforms
and brass-hats by the score.
Farewell! ye lights of Simla,
you'll lure me back no more.

For three long years of boredom
I've sat with sullen lips,
slave of red tape and telephones,
of flies and urgent slips.

Farewell to Simla

A dull quill-driving babu
with daily thinning hair,
girth swelling, liver torpid,
on a creaking office chair.

Now I've shaken off the shackles
and I am free to roam
with spear and gun and rifle,
and the saddle for my home.

Boar in the Ganges *Kadir*?
Kashmir and stag and bear?
or the tiger and the bison
in their Central India lair?

But which? the panting up-hill stalk
after a record head?
or black and yellow stripes that glide
up the white *nulla*-bed,

With my old 'four-fifty' cuddled
and every sense awake?
or *wuh jata* from the beaters
when they see a grey boar break?

All good; all spell the open air,
and sun, the jungle-side;
but after years on office stools,
give me a ride, a ride!

So, heigh for Kadir grass and *Jhow*!
for horses and for spears!
and Kadir boars to pay me for
the drudgery of years!

Farewell! farewell ambition!
I mock you now I'm free.
Farewell! dear Simla ladies,
you have no charms for me!

Kadir rough grass and tamarisk country bordering big rivers   *nul'a(h)* ditch, dry or wet watercourse   *wuh (woh) jata* 'there it goes'   *jhow (jhao)* tamarisk

Farewell! black coats and uniforms,
and brass-hats by the score.
Farewell! you lights of Simla,
you'll lure me back no more.

'Lamcoll'

Vol. V (1932): Sir Boaster and the Giant Pig

(dictated and illustrated by Quintin Ambler, son
of an Indian Cavalry officer, at just six years old)

'Once upon a time there was an officer called Colonel Sir Boaster in
Army Headquarters. He was very keen on pigsticking. One day when
he was asleep in his bed at Simla he had a dream about a great big
Giant Pig with huge tusks that was living at Mashobra with his family.
The Mother Pig said to the Father: 'Oh, let's tusk that man; he is very
cruel to us, and loves to stick his spears into us.' So when Sir Boaster
woke up he went to the Kermander and Chief to ask him if he could
go on leave. The Kermander and Chief said: 'You may go on leave
for a hundred and fifty weeks and then you must come back.'

So Sir Boaster went with his horse 'Tommy', his spears and
everything as fast as he could to Mashobra.

When he got near the pig's house he heard a 'Woof! Woof!' and
out came the giant pig and sent his horse flying and hit him down on
his nose and tusked him a little and then called to Mother and Babies
and they all tusked him a little. Then they went away and he got on
his pony with one leg off. So he rode side-saddle so that he couldn't
let anyone see his foot off.

The Kermander and Chief said when he saw him: 'What's the
matter, why are you riding side-saddle?' So Sir Boaster said: 'I have
been tusked by a Giant Pig and his family and my leg is off and I rode
side-saddle so that no-one could see.'

'Oh My!' said the Kermander and Chief: 'You shall be a scout for
being so brave' and he made him a wooden leg and sent him to join
the Boy Scout Army.'

Hunting Song

Pledge me woman's lovely face,
Beaming eye and bosom fair—
Every soft and winning grace,

Sweetly blended sparkles there.
Is there one whose sordid soul
Beauty's form hath ne'er adored?
From his cold lip dash the bowl,
Spurn him from the festal board.

Pledge me next the glorious chase,
When the mighty boar's ahead;
He, the noblest of the race,
In the mountain jungle bred,
Swifter than the slender deer
Bounding over Deccan's plain;
Who can stay his proud career,
Who can hope his tusks to gain?

Pledge me those who oft have won
Tusked trophies from the foe,
And in many a famous run,
Many a gallant hog laid low;
Who, on Peeplah's steepy height
And on Ganga's tangled shore,
Oft again will dare to fight
With the furious jungle boar.

                              Anon

(With acknowledgements to Theodore Oliver Douglas Dunn, Author
of *Poets of John Company*, 1921, Bodleian Library, 2805 d 78.)

The Water-Fiend

Said Ganga Bishn to Radha Krishn:
'Come along fishin', it is not far;
I'll bet you a *kauri*, the evening's showery
and the fish bite freely in broad Bisarh.'

Said Radha Krishn to Ganga Bishn:
'Your heart is wishin' the same as mine;
I've forgotten the look of a Limerick hook
and this is the weather to wet a line.'

Kauri shell currency

'But,' said Radha Krishn, 'on one condition;
we mustn't get caught by the evening star;
for a diving devil is known to revel
below the level of broad Bisarh.'

They have chosen their nooks and baited their hooks,
and the fish bite freely in broad Bisarh;
intent on their curry, they do not hurry,
when stealthily peeps the evening star!

'O Radha Krishn,' cried Ganga Bishn,
as out went a thousand yards of *sut*,
'as sure as fate he has gorged my bait
and I've hooked the Pandubaliya *Bhut*.

'O Radha Krishn, dear Radha Krishn,
lend me a helping hand I beg!
For the bank is steep, and the tank is deep,
and the line is twisted round my leg!'

Said Radha Krishn, 'With your permission,
I'd rather not stay 'neath the evening star;'
so he fled from the bank, and his poor friend sank,
and the Bhut had dinner in broad Bisarh.

T.F. Bignold, ICS, *Leviora*, 1963–88

Indian Golf

If links were sometimes grassy,
and Indian suns less hot,
then golf might be enticing.
If I could leave off slicing
with driver, spoon or brassy,
and play a decent shot,
if links were sometimes grassy
and Indian suns less hot.

Bisarh   reservoir near Gaya, avoided by locals as it is thought to be haunted
by a devil (*bhut*)   *sut* cotton thread   Pandubaliya   pandubali is one who is
the cause of others diving

If golf were more like hockey,
and I could hit the ball,
if 'browns' were much less bumpy,
and nerves were not so jumpy,
and fairways never rocky,
I'd not complain at all;
if golf were more like hockey
and I could hit the ball.

If cows did not distract us,
and drives required less aim,
if golf balls cost but little,
and clubs were not so brittle,
and courses free from cactus,
then golf might be a game;
if cows did not distract us
and drives required less aim.

'Momos'

The Sticking of the Pig

'Just the place for a Pig*!' said the Joint with a sneeze,
    as his pony sat down in a pool—
it was not; but on awkward occasions like these,
    it is always as well to keep cool.
'Just the place for a Pig!' Every man who could wield
    an arm, had turned out with a rush—
from the Master who wanted to write to the *Field*
    to the Parson who wanted 'the brush'.

They'd a District Surveyor with compass and maps
    a Telegraph Clerk, and a Tout,
and a Factory Expert, who thought that perhaps
    the run would be good for his gout.
The Judge, who was cautious, instead of a spear,

---

*Whenever mentioned, the wild boar is intended   the *Field* English magazine devoted to field sports, country houses, and gardens

was armed with a six-barrelled Colt,
to settle the Pig should it venture too near—
      or his horse, should it venture to bolt.

The last to arrive was a *Bacha*, who came
      on a charger he'd bought at a sale—
a single-eyed caster, inclined to be lame,
      and without any vestige of tail;
it had one wooden leg, and a ponderous paunch,
      and its hocks were excessively big,
but for all that, its owner maintained, it was 'staunch'—
      the great thing when you're sticking a Pig.

. . . .

'Just the place for a Pig!' so they formed up in lines,
      while the Joint who had been out before,
expounded the four unmistakable signs
      by which to distinguish a BOAR.
'Let us take them in order—the first is its size,
      for it's thirty-six inches in height,'
the District Surveyor here muttered a prayer,
      and all of them turned very white.

'The next is its tushes, which flash when it rushes
      and fizzle and hiss through the air;
and the third is its fondness for ripping you up,
      and not clawing you down, like a bear.
The last is its singular freedom from claws,
      which prevents it from climbing up trees.'
The last was received with a round of applause,
      as they tightened the girths of their 'gees'.

They had searched all the morning and finished their lunch,
      when the Master declared that he found
the presence of poppy productive of 'copy'
      and lay down to write—on the ground.

   Bacha (*bachcha*) literally child, also used to mean a young man, naive yet energetic

They had burnt all the jungle and poisoned the wells,
    and the Parson sat down on the brink
of a hole in the sand, which he cunningly planned
    to watch till the Pig came to drink.

They had harried the sugar with beaters and ropes,
    and the Judge, who had just had a spill,
said he'd sojourn awhile by the corpse of a *bail*,
    which he thought was a 'genuine kill'.
They were just on the point of concluding the hunt,
    as all of them felt very sore,
when from out of the 'jhao' came an ominous grunt,
    which they thought must proceed from a Boar.

'It's a Pig,' cried the Joint, as he held up his hand
    for a halt in the orthodox style—
'It's a Pig,' cried the Priest from his hole in the sand,
    cried the Judge from the corpse of his 'bail':
the Factory Expert had climbed up a tree,
    and the 'bacha', though still full of fight,
was heard to complain, when he pulled his left rein,
    that his charger would charge to the right;

though somewhat perplexed by this trick in his 'gee',
    he was greatly encouraged to find
though it certainly shied from the side it could see,
    it was staunch on the side it was blind.
The Pig's safest course was, no doubt, to lie still
    and remain unobserved by its foe,
but it rashly poked out an inquisitive snout,
    and a shot from the Judge laid it low.

They sprang like wild beasts on their moribund prey,
    which was stretched on its back in the 'jhao',
but they softly and silently galloped away,
    for the Pig after all was a Sow.

C.H.B. Kendall, ICS,
The Aged Joint and Other Verses, 1922

Bail buffalo

The Peora Hunt

Pit where the buffalo cooled his hide,
by the hot sun emptied, and blistered and dried;
log in the plume-grass, hidden and lone;
dam where the earth-rat's mounds are strown;
cave in the bank where the sly stream steals;
aloe that stabs at the belly and heels,
jump if you dare on a steed untried—
safer it is to go wide—go wide!
Hark from in front where the best men ride:
'Pull to the off, boys! Wide! Go wide!'

Kipling, *Plain Tales from the Hills*, 1888,
chapter heading to 'Cupid's Arrows'

Hunting Song (Bihar Light Horse)

Over the valley, over the level,
through the wild jungle we'll ride like the devil.
Hark! For'ard a boar! Away we go;
sit down and ride straight. Tally-ho! Tally-ho!

He's a true-bred 'un, none of your jinking;
straight across country, no time for thinking!
The nullah in front yawns deep as hell;
but a boar's gone through—we must go as well.

The ditches and banks are wide and steep,
the earth is rotten, the water deep;
the boldest horseman holds his breath,
but he must cross it to see the 'death'.

Over we go, the hunt's nearly done,
the field is gaining, the race is won;
an arm upraised, then a dash and a cheer,
and the boar has felt the deadly spear.

See how he flashes his fiery eye,
ready to charge, to cut and to die;
a boar that will charge like the light brigade
is the bravest beast that e'er was made.

Gentlemen, I won't detain you a minute,
I hope every glass has got something in it;
come fill them up with a bumper more;
are your glasses charged? Mr Vice—'The Boar'.

Chorus

Over the valley, over the level
through the wild jungle we'll ride like the devil;
there's a nullah in front and a boar as well;
sit down on your saddles and ride like hell.

Anon, *c.* 1856

5

Epigrams, Bon-Mots, Jingles, Eating and Drinking

On a Station in Lower Bengal

Our church as at present it stands
has no congregation, nor steeple;
the lands are all low-lying lands
and the people are low lying people.

   T.F. Bignold, ICS, *Leviora*, 1963–88

'A judge unrobe in public?' 'Yes, my brother,
you put a rent suit off and try another.'

                    Ibid.

Epigram VI–Truth

Bare falsehood found, as to and fro she hovered,
    truth bathing at the bottom of a well
and stole her clothes and semblance. Truth, uncovered,
    blushed to be seen abroad—so poets tell.
I longed for truth, and earnestly I sought her
    throughout the day, until I went to dine;
I found her then, not in a well of water,
    but at the bottom of a bowl of wine.

         A.G. Shirreff, ICS, *The Dilettante*, 1913

Roundel

Hammer, hammer, hammer, on the hard high road;
    solitary riding loses all its glamour
on the vilest mare that ever man bestrode—
jibbing when I fondle,—bucking when I lamm her.—

Poetry of motion? Poetry be blowed!
    'Tisn't even prose but gerund—grinding grammar—
        hammer, hammer, hammer.

Mare, your fate is sealed, your cup has overflowed;
    'Going, going, gone,' the auctioneer shall clamour;
Thrice he shall his desk, according to his code,
    Hammer, hammer, hammer.

                Idem, *Tales of the Sarai*, 1918

Mofussil Sunday

After a week of *jamabandi*,
I thank thee, Lord, for this thy Sunday
that mitigates the heat and stress
with a good Home Mail and a win at chess.

      Sir J.A. Thorne, ICS, *Madras Mail*, 1931

Motacilla Alba

Where have I known the like of these
wagtails, walking with dainty paces,
grey, neat and slim? Why, ADC's
all dressed up for the races!

                    Ibid.

Lady H. commenting on a man with rather fixed and rigid ideas, said:
'But what could you expect from the offspring of a couple who shared
the same bed for 21 years before he was born?'

                      Anon

The Old Men

This is our lot if we live so long and labour unto the end—
that we outlive the impatient years and much too patient friend:
and because we know we have breath in our mouth and think we
    have thought in our head,
we shall assume that we are alive, whereas we are really dead.

      Rudyard Kipling, *The Five Nations*, 1903

Jamabandi   annual check of revenue accounts in every village in every district

'For let who will rebuke or gibe,
my soul is numbered of that tribe
who nightly pitch their moving camps—
of pilgrims, vagabonds and tramps—
the Earth's and Nature's nearest kin,
who toil not neither do they spin—
yet Solomon with all his wealth,
had not their freedom, nor their health;
and most assuredly—ahem—
was not arrayed like one of them.'

Charles Russell, Principal of Patna College,
in Harry Hobbs, *Scoundrels and Scroungers*

'He slept beneath the moon.,
he basked beneath the sun,
he lived a life of going to do,
and died with nothing done.'

James Alberry, in Ibid.

'Stranger, these ashes were a man's,
crushed with a grievous weight;
he had acquired more ignorance
than he could assimilate.'

Idem, *Book of the World*

Easy communications make bad manners.

The sea's an excuse to be disgracefully loose in behaviour, voice and attire.

A soldier complained: 'What with the hum of mosquitoes above and the bugs in the bed below, I am regularly humbugged out of my night's rest'.

Ibid.

Jingles

There was an old man of Darjeeling
who kept his eyes fixed on the ceiling;
    no wonder he missed
    'Blue Peters' at whist
and made such a bungle of dealing.

A lady who came to Calcutta
was poisoned with clarified butter;
    'Oh Consomer *jee*,
    you have killed me with *ghee*!'
the very last words she could utter, or mutter,
that lady who came to Calcutta.

A little girl lived near a tank
with a half-witted father named Frank;
    they victualled a *sampan*
    with breadpan and *jam-pan*
and sailed for nine years till she sank.

        T.F. Bignold, ICS, *Leviora*, 1963–88

Limericks

There was a young lady of Cirencester,
O lucky the fellow who kissest her!
    So far it's been done
    by one lucky one
but then, you see, she is his si-sister.

There was an old Khan of Kalat,
who said to the Wali of Swat:
    ' 'Twas a very good shot,
    I admired it a lot
but I'd rather you used your own hat.'

There once was a ruler obscure,
who expected rebirth as a *chuha*.
    said, ' is a fallacy':
instead he constructed a sewer.

        R.V. Vernède, ICS, 1936

Consomer (*khansamah*) cook, anglicized version   *jee* suffix used in formal address   *ghee* clarified butter   *sampan* a kind of small boat   breadpan word invented by Bignold, to match with 'jam'pan   Khan and Wali titles of the rulers of Kalat and Swat respectively, two small NW Frontier states   *jam-pan* a kind of sedan chair carried by four persons, in hill stations   *chuha* rat

There was nothing particularly attractive about life in Calcutta fifty years ago. The drinking water was unpleasant and unwholesome. The air of the city was permeated with a disgusting odour. Side streets were sodden with the filth of generations and stank like hospital ship clothing packed in the Red Sea. In spite of general complaints, Calcutta under European management was dirtier in every respect than it is to-day.

. . . .

It must be admitted there were few hypocritical refinements about our daily life. Goat-face teetotallers or those who had jobs as missionaries and shuffled about, flat-footed, arrogantly pure in heart, while as mean as maggots, would have been as much objects of our contempt as a man caught reading poetry. Our opinions were crude—but candid.

. . . .

We took the world as we found it and, as much as possible on our pay, enjoyed ourselves. Discontent does not go deep when none are rich. It is comparisons which make life odious.

. . . .

Everything in life has its opposite number and there are drawbacks to permanent jobs. They have a tendency to dope energy and to develop idiosyncrasies. Few will deny that in an international contest for cranks and eccentrics, drunk or sober, British Indians, if placed second, would be justified in lodging a protest. But how they add to the fun of life! Cranks drunk and eccentrics sober have left an indelible record all over the East.

Harry Hobbs, *Our Grief Champion*

Postage to England in the 1880s for a letter weighing half a *tola* cost four *anna*s. That made fashionable a thin transparent highly glazed notepaper, the produce of France, on which most people wrote with a treacly violet ink. Travellers carried their ink in a safety bottle, which, like fountain pens, often leaked. But the safety pen has driven the safety ink bottle off the market. Before posting letters it was customary to deface the stamps by drawing two lines with a pen across them. . . .

---

1 *tola* = 10 grams   16 *anna*s = 1 rupee

When this was declared illegal, it was felt that no letter would ever reach its destination, because the stamps would be stolen.

Idem, *Indian Dust Devils*

'Oh, Dick, you may talk of your writing and reading,
your logic and Greek, but there's nothing like eating.'

. . . .

Everybody did their best to put a square meal in a round hole, believing that a little is a good thing but too much is enough.

. . . .

One of those women who believe that food is what you have between meals, told a few sympathizers who called to see how she was: 'I've had my diet, now I'll have my dinner,' and put things right by declaring: 'I eats well, but I ain't gorgeous.'

. . . .

Count de Warren's (*c.* 1800) opinion of the manners of these ladies is not very favourable. He admitted that they were more intelligent than ladies of the same class in France but complains that they affected a childish simplicity. [He expected that, having been brought up on the Bible, they would know the facts of life. But the simplicity was probably adopted as a precaution against the wiles of the Frenchman—Ed.]

But now we have the young ladies at dinner. If you are a Frenchman you will be thunderstruck at the enormous quantity of beer and wine absorbed by these English ladies, in appearance so pale and delicate. I could scarcely recover from my astonishment at seeing my fair neighbour quickly dispose of a bottle and a half of very strong beer, eked out with a fair allowance of claret, and wind up with five or six glasses of light but spirited champagne, taken with her dessert. The only effect it seemed to produce upon her was visible in the diminished languor of her manner and the increased brilliancy of her eyes.

. . . .

'Quiz', an anonymous character in a publication called 'Que Hai' (1816), expresses disgust at seeing one of the prettiest girls put away two lbs of mutton chops at one sitting.

. . . .

George Jhonson (1830) wrote: 'When a lady is challenged during dinner, she very frequently takes beer instead of wine. . . . I have heard of four 'burra bebbees' who in the olden time daily took tiffin in each other's houses and drank a dozen of Hodgson's pale ale before they retired from the table to their couches . . . and I watched a lady after dinner put away six quarts of Allsopps without moving from her chair.

Idem, *John Barleycorn Bahadur*

[Reading such passages from *John Barleycorn Bahadur*, one wonders if such over-eating and drinking was general in Calcutta in the eighteenth and early nineteenth centuries. If so, did it reflect something peculiar to Calcutta and the life lived there, or merely the fashions prevalent in England at that time? Whatever the reason or reasons, such habits must have contributed significantly to the early deaths of so many young people. As John Barleycorn remarks: 'But then, as now, men did not die—they killed themselves; the difference being that in the good old days they did not take so long about it'—Ed.]

Fashion in Men's Clothes

When I arrived in India some two and twenty years ago, the extreme heat of the climate, ease and convenience seemed to lead the fashion; being principally considered with respect to dress, in preference to the reigning mode of a country where frost and snow, or cold damp weather, predominates for more than half the year. A silk coat, though cheap and easy to procure, was not on that account despised; nor was a profusion of powder and pomatum considered as a necessary lid for the head. Even the light garb before-mentioned (sili coat) was at convivial meetings laid aside and, stripped to their sleeve waistcoats, the young Writer and old senior merchant agreed in allowing themselves every decent chance of coolness and refreshment.

But alas! *tempora mutantur*, fashion in everything bears sovereign sway, and cloth coats, with hair dressed *à la mode d'Europe*, are now in vogue, the long buckskin meets the short boot, whilst the strings of the one and the straps of the other add not less weight than ornament to the knees and ankles. Even the gingham waistcoats, which, stripped or plain, have so long stood their ground must, I hear, ultimately give way to the stronger Kerseymere, of which I have seen complete suits worn in the hottest weather. No longer a stock or anything like a cravat, but a monstrous roll of stuffed muslin surrounds the neck; and the more effectively to guard against sore throats, that again is encompassed by the coat and waistcoat collars of modern magnitude. Now, as I am by no means a convert to the Spanish maxim that 'What keeps out the cold will keep out heat' such dress appears to be better adapted to the frozen climate of the Orkneys than the warm latitude of 13° 11.

The unaccommodating disposition of the English with respect to weather has, I find, been remarked by natives of the coldest climate of Europe, who have expressed great astonishment at seeing them exchange a warm room for a bleak atmosphere in the midst of winter without any precaution to guard against the cold. I have merely quoted these [examples] as proof of our English singularity; to show that of all the people under the sun our countrymen alone, as if insensible to cold or heat, affect to brave all vicissitudes of climate and obstinately to wear the same dress in every part of the world.

Hugh Boyd, in *The Indian Observer*, no. XIV,
10 December 1793

(Cf. Kipling in another context: 'For Allah created the English mad—the maddest of all mankind.' 'Kitchener's School', 1898, in *The Five Nations*, 1903)

John Collins, Billy Williams, Elsie May

All drinks are good in Hindostan,
but some there are which tumble slicker
down the parched throat of thirsting man
than other rival kinds of liquor:
and three there are whose names contain
hints of mysterious romances
which lead the thoughtful poet's brain
to weave anew delightful fancies.

'John Collins' forward! Who were you
that an immortal fame could win
by such a sweet innocuous brew
as bitters, lemon squash and gin?
Whence came that deep tremendous thought,
amounting to an inspiration,
famed as the apple-fall which brought
Newton ideas of gravitation?

Next 'Billy Williams'! Why was thine
the luck which led thee to dilute
thy lemon squash with ginger wine?
Why immortality the fruit
of what the simple person thinks
a strangely commonplace invention,
deserving among other drinks
merely an honourable mention?

Last 'Elsie May'! Now though to me
such prejudice is out-of-dated,
still 'tis unusual to see
ladies with drinks associated,
so read the riddle right I pray,
and, truth from fiction to dissever,
tell why the name of Elsie May
rings through the bars of time for ever?

I see a maid with flashing eyes
who, to a crowd of swains demented,
proclaimed her heart and hand the prize
of him who some new drink invented.
Art in simplicity we find,
and so the gentle maid bestowed her
sweet self on him whose taste combined
bitters, a lime and whisky soda.

Their spirits linger round the bar
as spirits have a way of doing,
and often have I seen from far
a sparkling 'John' or 'Bill' pursuing

a coy young 'Elsie' down the path
where spirits are too fond of going—
though what might be the aftermath
I have no earthly means of knowing.

Alas! those times are past and done;
the age of dulness is upon us;
and no-one nowadays has won
those high eponymistic honours,
or earned the meed of frank applause,
the burst of thrilling admiration,
paid by our tippling ancestors
to authors of some new potation.

And yet a dream there comes to me—
call it not fatuous or silly—
that I, some day, may numbered be
with 'Elsie May', and 'John' and 'Billy';
that I may conjure up my soul,
seize with a vital hand the squeegee,
and, filling high my brimming bowl,
may nickname some new drink 'R.G.G.'.

R.G. Gordon, ICS
(Brother of Jan Gordon), Bombay 1917

Second Fiddle

My father was a second son, a penniless cadet;
my mother was his second wife and I her second pet;
our flat was on the second floor, our wines a second brand;
our friends were rather second-rate, our fittings second-hand.

I missed (it was my second shot), I must have been a fool,
the second exhibition at a secondary school;
The Green-eyed Monster stung me then but time has drawn his sting;
It's second nature now with me to be the second string.

I faced the disappointment and, resolved to persevere,
was second in the second form within my second year;
of 'proximes' and 'mentions' I accumulated trucks,
but never won a single prize; I wasn't born a dux.

At games and sports it was the same; I entered for a race
and, when I got my second wind, was sure of second place;
at golf I lost my second ball before the second green;
at cricket if I stopped the first, the second bowled me clean.

I fancied second helpings but, when tempted to exceed,
on second thoughts declined them as I knew they disagreed.
I travelled second/season (there were seconds in those days)
and figured as a second in innumerable frays.

I wooed a second cousin but her mind was set on dross;
she said she might consider me perhaps *en secondes noces*
and hoped I'd grave her wedding ('twas a stony-hearted jest)
as the lucky bridegroom's best man and the sweet bride's second best.

I joined a Regiment of the Line; a soldier's life was grand,
but, when promotion ceased for me at second in command,
I thought I'd make a second start before it was too late,
so chose the mercantile marine and rose to second mate.

I'm now in second childhood and my second teeth are thin;
my second sight is failing and I've grown a second chin;
it is time I was seconded for a place upon the shelf,
and, if someone will propose it, I will second it myself.

A.H. Vernède, ICS, 1926

6
Birds, Insects, Animals

Insects

Alas! creative nature calls to light
Myriads of winged forms in sportive flight,
When gathered clouds with ceaseless fury pour
A constant deluge in the rushing shower,
On every dish the booming beetle falls,
The cockroach plays, or caterpillar crawls;
A thousand shapes of variegated hues
Parade the table, and inspect the stews!
To living walls the swarming hundreds stick,
Or court a dainty meal, the oily wick;
Heaps over heaps their slimy bodies drench,
Out go the lamps with suffocating stench!
When hideous insects every plate defile,
The laugh how empty and how forced the smile!
The knife and fork a quiet moment steal,
Slumber secure, and bless the idle meal;
The pensive master, leaning on his chair,
With manly patience mutters in despair—
O England! show, with all thy fabled bliss,
One scene of real happiness like this!

Anon

(With acknowledgements to Theodore Oliver Douglas Dunn, author
of *Poets of John Company*, 1921, Bodleian Library, 2805 d 78.)

Divided Destinies

It was an artless *Bandar*, and he danced upon a pine,
and much I wondered how he lived, and where the beast might dine,
and many other things, till o'er my morning smoke,
I slept the sleep of idleness and dreamt that Bandar spoke.

Bandar monkey

He said: 'O man of many clothes! sad crawler on the Hills!
Observe I know not Ranken's shop, nor Ranken's monthly bills!
I take no heed to trousers or the coats that you call dress;
nor am I plagued with little cards for little drinks at Mess.

I steal the *bunnia*'s grain at morn, at noon and eventide,
for he is fat and I am spare. I roam the mountain-side,
I follow no Man's carriage and no, never in my life
have I flirted at Peliti's with another Bandar's wife.

O man of futile fopperies—unnecessary wraps;
I own no ponies in the hills, I drive no tall-wheeled traps.
I buy me not twelve-buttoned gloves, 'short-sixes' eke, or rings,
nor do I waste at Hamilton's my wealth on 'pretty things'.

I quarrel with my wife at home, we never fight abroad;
but Mrs B. has grasped the fact I am her only lord.
I never heard of fever—dumps nor debts depress my soul;
and I pity and despise you! Here he pouched my breakfast-roll.

His hide was very mangy and his face was very red,
and ever and anon he scratched with energy his head.
His manners were not always nice, but how my spirit cried
to be an artless Bandar loose upon the mountain-side!

So I answered: 'Gentle Bandar, an inscrutable Decree,
makes thee a gleesome fleasome Thou and me a wretched Me.
Go! Depart in peace, my brother, to thy home amid the pine;
yet, forget not once a mortal wished to change his lot with thine!'

Rudyard Kipling, *Departmental Ditties*, 1886

The next four articles are taken from *In My Indian Garden*
by Phil Robinson, United Provinces, 1888

Green Parrots

The crow requires much time to develop and perfect his misdemeanours; the parrot brings his mischiefs to market in the green leaf. The first is a crafty calculating villain; the latter is a headlong blackguard. While a crow will spend a week with a view to the ultimate abstraction of a key, the parrot will have scrambled and screeched in

---

Bunnia Indian merchant

a day through a cycle of larcenous gluttonies, and before the crow has finished reconnoitring the gardener, the parrot has stripped the fruit-tree. We do have surely a reasonable cause for complaint, when nature creates thieves and then gives them a passport to impunity. For the green parrot has a large brain (some naturalists would like to see the Psittacid family on this account rank first among birds), and he knows that he is green as well as we do, and knowing it he makes the most of nature's injudicious gift. He settles with a screech among your mangoes, and as you approach, the phud! of the falling fruitlings assures you that he has not gone. But where is he? Somewhere in the tree you may be sure, probably with an unripe fruit in his claw, which is raised halfway to his beak, but certainly with a round black eye fixed on you, for, while you are straining to distinguish green feathers from green leaves, he breaks with a sudden rush through the foliage, on the other side of the tree, and is off in an apotheosis of screech to his watch-tower on a distant tree. To give the parrot his due, however, we must remember that he did not choose his own colour—it was thrust upon him; and we must further allow that, snob as he is, he possesses certain manly virtues. He is wanting in neither personal courage, assurance, nor promptitude, but he abuses these virtues by using them in the service of vice. Moreover, he is a glutton, and, unlike his neighbours, the needle of his thoughts and endeavours always points towards his stomach. Not only is his appetite inordinate, but his wastefulness also, and what he cannot eat he destroys. He enters a tree of fruit as the Visigoths entered a building. His motto is 'What I cannot take I will not leave', and he pillages the branches, gutting them of even their unripest fruit. Dr Jerdon in his *Birds of India* records that 'owls attack these birds at night', and there is, ill-feeling apart, certainly something very comfortable in the knowledge that while we are warm a-bed, owls are most probably garrotting the green parrots.

The Seven Sisters (*The Babbler Thrushes* Malacocircus)

As busy as the Mynas, but less silent in their working, are these sad-coloured birds, hopping about in the dust and incessantly talking while they hop. They are called by the natives 'The Seven Sisters', and seem to have always some little difference on hand to settle. Fighting? Not at all; do not be misled by the tone of voice. That heptachord clamour is not the expression of any strong feelings. It is only a way they have. They always exchange their commonplaces as if their next neighbour was out of hearing. If they could but be quiet they might pass for the bankers among the birds. They look so very respectable. But though they dress so soberly, their behaviour is unseemly. Pythagoreans may, if they will, aver that these birds are the original masons and hodmen of Babel, but I would rather believe that in a former state they were old Hindu women, garrulous and addicted to raking about amongst rubbish heaps, as all old native women seem to be. The Seven Sisters pretend to feed on insects, but that is only when they cannot get peas. Look at them now! The whole family, a septemvirate of sin, among your marrow-fat peas, gobbling and gabbling. And it is of no use to expel them—they will return. When it is night they will go off with a great deal of preliminary talk to their respective boarding-houses, for these birds, though at times as quarrelsome as Sumatrans during the pepper harvest, are sociable and lodge together. The weak point of this arrangement is that often a bird—perhaps the middle one of a long row of closely packed snoozers—has a bad dream, or loses his balance, and instantly the shock flashes along the line. The whole dormitory blazes up at once with indignation, and much bad language is bandied about promiscuously in the dark. The abusive shower at length slackens, and querulous monosyllables and indistinct animal noises take the place of the septemfluous (Fuller has sanctified the word) vituperation, when some individual, tardily exasperate at the unseemly din, lifts up his voice in remonstrance, and rekindles the smouldering fire. Sometimes he suddenly breaks off—suggesting to a listener the idea that his next neighbour had silently kicked him—but more often the mischief is irreparable, and the din runs its course, again dwindles away, and is again relit, perhaps more than once, before all heads are safely again under wing.

The Sparrow

'The sparrow,' said Luther, 'is a most voracious animal, and does great harm to the crops. The Hebrews call it "tschirp", and it should be killed wherever found.' It is Emile Souvestre who calls the sparrow 'the nightingale of the roofs', and says that 'our chimney-pots are his forests, and our slates his grass plots', but I incline to take a less lenient view of the genus *Passer*. As we resent the likeness to ourselves which the monkey tribe possess, so we feel injured by the familiar communism of the sparrow. He professes to move on the same plane with man—our chimney-pots are his chimney-pots and our slates his slates, but our forests and grass-plots are none the less his also. There is in his deportment none of the deference of a stranger when he crosses your threshold. His entry is that of a conqueror into a hostile city. He begins by putting himself on an equality with you, but soon arrogates superiority. He holds that man by natural selection will develop into the sparrow, but in his present hybrid state criticizes him as the fool who builds houses for the wise sparrow. Show me a man's house and I will show you a sparrow's castle; point out, if you can, a stable which the sparrow does not share with the horses.

He is the gamin of birds—chief vagabond of the air. He it is who crowds without payment into places of public amusement, disturbs Divine Service by a fracas with his kind on the altar rails, or perches above the Commandments and chirps monotonously through the sermon. His cranial development is very poor—flat atop, showing a deplorable lack of respect; bulgy behind, typical of gross amativeness and gluttony; and puffy at the sides where lodge the devils of destructiveness, evil-speaking, lying, and slandering. This Bohemian communist has broken through—worn out—the resentment of man; we no longer resist his intrusions or retaliate for his rapine. He has acquired a prescriptive right to be iniquitous and go unpunished. But he does not understand this. In his conceit he insolently imagines that he has compelled acquiescence and treats us as a conquered race. In another world he will be me met with strolling in the valley of Jehoshaphat, flower in hand—the badge of one who has benefited his fellow-men—will swagger through the fields of amaranth and moly, and take to himself more than his share of asphodel.

The Mynas

There are undoubtedly among the feathered race some to whom a Republic would present itself as the more perfect form of government, and to none more certainly than to the Mynas. The myna is, although a moderate, a very decided republican, for sober in mind as well as in apparel, he sets his face against such vain frivolities as the tumbling of pigeons, the meretricious dancing of peafowl, and the gaudy bedizenment of the minivets, holding that life is real, life is earnest, and, while worms are to be found beneath the grass, to be spent in serious work. The myna, therefore, views with some displeasure the dilettante hawking of bee-eaters and the leisurely deportment of the crow-pheasant; cannot be brought to see the utility of the luxurious hoopoe's crest, and loses all patience with the koel-cuckoo for his idle habit of spending his forenoons in tuning his voice. For the patient kingfisher he entertains a moderate respect, and he holds in esteem the industrious woodpecker; but the scapegrace parrot is an abomination to him and, had he the power, he would altogether exterminate the race of humming-birds for their trifling over lilies. Life with him is all work. Of course he has a wife, and she celebrates each anniversary of spring by presenting him with a nestful of young mynas, but her company rather subdues and sobers him than makes him frivolous or giddy, for as the myna is, his wife is—of one complexion of feather and mind. A pair of mynas remind one of a Dutch burgher and his frau. They are comfortably dressed, well fed, of a grave deportment, and so respectable that scandal hesitates to whisper their name. In the empty babble of the Seven Sisters, the fruitless controversies of finches, the bickering of amatory sparrows (every sparrow is at heart a rake), or the turmoil of kites, they take no part—holding aloof alike from the monarchical exclusiveness of the jealous *Raptores* and the democrat communism of the crows. The myna shrinks from the neighbourhood of the strong and resents the companionship of the humble. He comes of a race of poor antecedents, and he has no lineage worth boasting of. The crow has Greek and Latin memories, and for the antiquity of the sparrow we have the testimony of Holy Writ. It is true that in the stories of India the myna has frequent and honourable mention; but the authors speak of the hill-bird—a notable fowl with strange powers of mimicry, and always a favourite with the people—and not the homely Quaker bird who so diligently searches our grass-plots, and may be seen from dawn to twilight, busy at his appointed work—the consumption of little grubs. The lust of the green parrot for orchard brigandage, or of the proud-stomached king-crow for battle with his kind, are as whimsical caprices, fancies of the moment, when compared to the steady assiduity with which this Puritan bird pursues the object of his creation. And the result is that the myna has no wit. Intelligence is his of a high order, for busy as he may be, the myna descries before all others the far-away speck in the sky which will grow into a hawk, and it is from the myna's cry of alarm that the garden becomes first aware of the danger that is approaching. But wit he has none. His only way of catching a worm is to lay hold of its tail and pull it out of its hole—generally breaking it in the middle, and losing the bigger half. At night the mynas socially congregate together; and, with a clamour quite unbecoming their character, make their arrangements for the night, contending for an absolute equality even in sleep.

Has it ever struck you how fortunate it is for the world of birds that of the twenty-four hours some are passed in darkness? And yet without the protection of night the earth would be assuredly depopulated of small birds, and the despots, whom the mynas detest, would be left alone to contest in internecine conflict the dominion of the air.

The next five articles are taken from *The Tribes on my Frontier* by E.H. Aitken, Bombay 1878–80

The Mosquito (July)

What possible gain can it be to a mosquito to gorge itself on my life-blood until its wings almost refuse to carry it, and it can just sail slowly, like some great crimson balloon, with the wind, positively inviting me to imbrue my hands in my own blood, and so avenge the wrongs of countless nights of woe? Insects, as everyone knows, or ought to know, require no food in their winged state—at least the flimsier kinds do not, such as flies and gnats, and butterflies. They have done all the serious business of life, the eating and growing, in their grub state, and when they dress up and come out into world, to enjoy a few days of vanity before they die, they have no proper mouths, only a sort of tube for sipping light refreshments. Supposing that mosquitos do require nourishing food why can they not bleed us painlessly? Why make us pay fees in anguish for the operation? It can be no advantage to them that we wince and jump when they sit down to dine. The traveller who invented the original vampire bat understood matters better and made the horrid monster fan its victim gently with its ample wings, that he might the more sweetly sleep on into the sleep of death. So, from the Darwinian standpoint, mosquitos ought to have developed some sweet narcotic fluid, some natural rosalpinus, which would produce the most exquisitely pleasant titillations, and make the fat man hasten to resign his back, sore vexed with prickly heat, to their soothing ministrations, and his soul to sweetest dreams. I hold that

Darwin, weighed in the balance against the mosquito, is found wanting.

To come to the description and history of the animal, the mosquito is not the same as the buffalo, though it is said that a young lady who had just landed in India fled from a herd of those peaceful domestic antediluvians and asked if they were not the dreadful mosquitos of which she had heard such tales. The mosquito is only a little insect with two wings and six legs. The wings are for flying with, and four of the legs for walking with. The two long hind legs are connected with the suction apparatus and are of the nature of pump-handles. Of course, the anatomist, prying with his microscope, will deny this; but the microscope comes from *micros*, small, and *scopein*, to see, and no one who relies on it can grasp a large idea. Anybody may satisfy himself by watching a mosquito looked at microscopically, contains a whole set of surgical instruments: looked at large-mindedly, it is simply the tube of an artesian well, and is used in the same way. When a mosquito settles on you it pricks up its ears for a moment, to make sure that there is no danger near, and then walks about slowly, probing for a soft place. When it has found one, it fixes the tube and begins to drive it home. Then is the moment to smite it.

Mosquitos are of many sorts. There are common grey ones; and small, speckled, shrill-voiced ones which sing an overture and then tap the outside edge of your ear; and large droning ones, which are found, like the best mangoes, only in Mazagon and some other parts of Bombay; and queer ashy ones, which stand on their heads and bore into you like a bradawl.

As to its history, all the 'promise and potency' of the future mosquito lay at first in a minute egg floating on dirty water. From this came forth an execrable shape, bristling all over with hairs, breathing through its tail, and progressing by a series of wriggles, bringing its head and tail together first on one side, and then, with a jerk, on the other. So, by making ends meet, it twisted itself through life for a fortnight or more, feeding day and night on the impurities of the water and growing prodigiously. Then it floated for a while, eating nothing but meditating a change. When at last internal arrangements were completed, the skin at the back of its head split open, and the mosquito looked out, snuffed the fresh air, drew itself cautiously out of its case, and glided gaily over the water on a boat made of its own skin. Then it sailed away into the air and joined the throng

As thick and numberless
as motes that people the sunbeams,
or likest hovering dreams.

Then, as dawn began to light up the eastern sky, they swarmed in through every open window, and took shelter among the folds of hanging coats, inside boots, in the pocket of the dressing-gown, in the chambers of the sola topee; and there they are! And what is to be done?

Well, by dusting and sweeping, and burning incense and folding all hanging clothes, you can make them very unhappy; and, for your own protection, you can make yourself utterly abominable to them by anointing your hands and face with toilet-water, or even eau-de-Cologne. But it is clear that the thing to do would be to come upon the sanguinary hordes in their earlier stages, and nip them in the bud, cut them off while they are only mosquitos *in posse*, not *in esse*. And this can be done, for, when a house is much plagued with them, it may be set down for certain that there is a factory on the premises. The first thing to do, then, is to make a tour of inspection. Go to the back of the kitchen and see if there is not a small cistern, or a tub sunk in the ground, connected by a small pipe through the wall with the arena of all Domingo's professional operations, a veritable Dead Sea, where baleful streams run in, but nothing runs out. There in the inky fluid, on which a filmy scum floats, whose rainbow radiance is broken only by the spluttering of the bills of happy ducks, you will find them in writhing swarms, sixteen to the superficial inch, fast ripening towards malefaction; and you may spill their lives not by tens or hundreds, but by quarts and gallons.

But all means of prevention are more or less disappointing, for after all it is ordained that mosquitos shall bite us. What is wanted, then, is some cure, or antidote, for the bite—and there is only one, of which I am the original discoverer. A bigoted old Brahmin, who never tired of unmasking the inherent badness of everything English, once admitted to me in a moment of candour that in one point we were better than his countrymen. 'If a Hindoo,' he said, 'invents or discovers anything, he keeps it a secret and makes all the profit he can out of it, and when he dies, it dies with him; but if the Englishman makes a discovery, he publishes it and the world gets the benefit.' So I will divulge my antidote for mosquito bites. It is inoculation. The idea is curiously supported by analogy, for Dr E Nicholson, in his book on snakes, speaking of the confidence with which Burmese snake-charmers handle the terrible *Ophiophagus elaps*, says that they certainly have some remedy, and he believes it is simply gradual inoculation with cobra poison. Such experience as we have points in the same direction. The griffin gets up in the morning with his face like a graveyard, a monument for every bite; but as his blood becomes accustomed to the poison, these violent effects cease. Probably the remedy has never been fully tried, but its success is certain. So, if anyone is much tormented by mosquitos, all he has to do is to dispense with curtains and let them bite him freely for a year, or two or three years (I am not certain how long it will take), until his constitution becomes mosquito-proof, and then for the rest of his days he may defy the most trumpet-tongued and asp-envenomed of the blood-thirsty race.

The Lizards (August)

One peculiar feature of life in India is the way we are beset by lizards, and nobody seems to notice it. We all come out to this country more or less prepared to find scorpions in our slippers, snakes twined about our hair, and white ants eating up the bed in one night, so that in the morning we are lying on the floor; but nobody warns us to expect red-throated hobgoblins clambering about the trellis, and snaky green lizards prying about the verandah at noon-day, and little geckos visiting the dinner-table at night. Perhaps, because they are not very pestilent enemies, nor very useful friends, shallow-minded people do not think them worth notice. But a contemplative spirit feels that it would starve without many things which are of no use in the gross sense of the word, and there is much matter for chastening meditation in lizards. Has not the poet said:

Men are we, and must grieve when even the shade
of that which once was great has passed away.

and lizards once were great. They were the aristocracy of the earth. What a strange world there must have been on this same earth of ours in those days! Did mosquitos as large as sparrows, with voices like tin trumpets, infest the swampy wastes and torment the drowsy megalosaurus, and did the winged lizards, like flying foxes, hawk them in the dusky forest? Did the mild iguanodon, when it had done browsing on a tuft of maidenhair fern about the size, say, of a clump of bamboos, turn round and waddle away into a hole, as its successors do to-day on the plains of Guzerat?

The lizards are the wreck of a great past. They had their day; perhaps they abused it; at any rate the great unresting wheel has gone round, and that which was up is down. The commonalty do not seem to feel it much. They have parted even with pride and make the most of their circumstances. But all the descendants of great families, the crocodiles and alligators and even iguanas, are a prey to melancholy. They maintain a dignified spiritlessness which is affecting.

The iguana or *Gorpud* has a name in history, I mean the history of man. The old tutelary Brahmin of Singhur, if he is still alive, delights to show sahibs the spot where the Marathas tied a strong light rope round the loins of a huge gorpud, and waited till it had clambered up the rocky face of the fortress and wedged itself into some rugged fissure; then, while it clung as they can cling, one sinewy mountaineer after another bound his waistcloth more tightly round him and climbed the rope in silence, laughing in his sleeve at the astonishment in store for the vigilant Mussulman garrison.

Like all races whose greatness is a memory, lizards are sensual, passionate and cruel. Sensual first: a lizard lives to eat, and there never seems to be any time in its life when it is not looking for food. And passionate next. Two sparrows will squabble and scuffle until they get so inextricably mixed that, when they separate, it is quite an open question whether they have got their own legs and wings, or each other's; and two ants will fight until they die in each other's jaws, and a third comes up and carries off the whole jumble for the food of the community; but for an example of devouring rage go to the big garden lizard, which the children of India call a blood-sucker. See it standing in the middle of the road, its whole face and throat crimson with wrath and swollen to the bursting-point with pent up choler, its eyebrows raised, and its odious head bobbing up and down in menace of vengeance. And the explanation of the whole matter is that another smaller lizard snapped an ant on which it had set its heart. Nothing will appease it now but to bite off the offender's tail. This will do the latter no harm, for a lizard's tail is a contrivance for the saving of its life, planned on exactly the same principle as the faithful Russian slave who threw himself to the wolves that were pursuing his master's sledge. I once saw a fierce scorpion catch a lizard by the tail and plunge its sting into the wriggling member; but before the venom could circulate to the lizard's body, it detached its tail and ran away grinning. The scorpion went on killing the old tail, and the lizard began growing a new one.

This was one of those little house lizards, called geckos, which have pellets at the ends of their toes. They are not repulsive brutes, like the garden lizard, and I am always on good terms with them. They have full liberty to make use of my house, for which they seem grateful, and say chuck, chuck, chuck. They are low-minded little plebeians, no doubt, and can see nothing in a satin-white moth with vermilion trimmings except wholesome victuals; but one must put up with that, for they do good service. At this season, when the buzzing pestilence of beetles and bugs is on us, they tend towards *embonpoint*, but they bate nothing of their energy, nor seem to get near the limits of their capacity. They hold that the Bombay Gas Company was established for their accommodation, and there is scarcely a gas lamp but has its guardian gecko, fat with moths and mantises, dragon-flies, grasshoppers, crickets and cockroaches, even hard-shelled beetles, but not blister-beetles. These would irritate their little insides, for the sake of which alone they live.

The only genteel member of the family is the green lizard. Its manners are graceful and unassuming, and its external appearance is always in harmony with the best taste, while it does not betray that ceaseless hankering for provisions which stamps the rest of them. It is timid and retiring, but as the sun grows hot in the forenoon you will hear it rustling among the leaves (*virides rubum dimovere lacertae*), then it will come softly up the steps, behind the *calladium* pots and along the wall of the verandah, and perhaps, if you keep very quiet, into the drawing-room. It does little good, eats a few ants, perhaps, but it enjoys itself and does no harm, and I have always had a leaning towards the green lizard.

I do not know whether I should class the chameleon among my frontier tribes, for the only one about my territories was born near Ahmednugger and is a state prisoner with me. His residence is a canary cage with green muslin all round to keep in the flies which I provide for his maintenance. Here, clutching a twig, as if he were the fruit that grew on it, he lives his strange life of motionless meditation. Till a late hour in the morning he sleeps, sounder than a *chowkeydar*;

Chowkeydar watchman

nothing will wake him. At this time his hue is a watery greenish yellow. When the sun begins to warm the world, then colour slowly comes back to his reviving limbs, and he appears in a dark earthy brown.

Through the day this is his livery, varied sometimes with specks of white and sometimes with streaks; but when the afternoon shoots its slanting rays through the bars of his cage, surrounding him with checkered light and shade, then he catches the same thought and comes out in vivid green with leopard spots upon his sides. Then, when night comes on, the same deathlike paleness again overspreads his torpid frame. Philosopher as he is, the chameleon requires food, and since he is too slow to go after it, he brings it to him. As his ball-and-socket eyes roll this way and that way, one of them marks a large white butterfly walking up the bars of his cage, and he forms a purpose to eat it. He unwinds his tail, then relaxes the grasp of his broad palms one at a time (for he is extremely nervous about falling and breaking his bones), and so he advances slowly along the twigs until he is within six inches of his prey. Then he stops, and there is a working in his swollen throat; he is gumming his tongue. At last he leans forward, and opens his preposterous mouth, and that member protrudes like a goose-quill steeped in white bird-lime. For a moment he takes aim, and then, too quick for the eye to follow it, the horrid instrument has darted forth, and returned like elastic to its place, and the gay butterfly is being crunched and swallowed as fast as anything can be swallowed when tongue, jaws and throat are smeared with viscid slime. But this part of the process is inconceivably vulgar, and we may well leave the chameleon to himself till it is over.

The Ants (August)

A glory battle has been fought in the bathroom, and the field of carnage is appalling to look upon. For some days past, curious, crabbed-looking, reddish-brown ants have been gathering in a lump about the mouth of a small hole in the floor. This means always that a new colony is to be founded. I have no objection to colonies in the abstract, but to see a teacup-ful of crusty little brutes heaped up on the floor not a yard from your tub has a tendency to make you feel uneasy, so I endeavoured to discourage them by dashing the 'tin pot' full of water at them and sweeping the whole body away in a flood.

But anyone who engages in a battle of obstinacy with ants should practise to suffer defeat gracefully, for he will have to suffer it. They put me to the trouble of keeping up this tin-pot practice for three or four days, into the jaws of destruction, they were back again in an hour with a few more. At length the emigrants appeared, great lubberly things, fully an inch and a half long, with wings, and not a notion of how to use them. The room was soon full of them, crawling over each other, or making blundering essays at aeronautics, which invariably ended in a butt against the wall. This brought on a fit of brain-fever, in which they spun on their heads like teetotums, or went sliding with a buzz-z-z! along the floor. Then the squirrels got scent of the affair and came in to munch them up, and the lizards swallowed them, and the sweeper swept the residue out to the chickens. So the colonizing scheme collapsed. To return, however, to my story. There is in the same room a settlement of those large black ants which come into the house at this season and garrison cool damp corners. They are truculent, hot-blooded ruffians, and will stomach no provocation, so it is little wonder that the two parties came into collision, especially at a time of such national excitement as always attends the ceremony of seeing an emigrant party off. The battle began in the evening, and I was there as a special correspondent for the World. The black ants were few in number, but terrible in their onslaught. They fought singly. I watched in particular one of gigantic build and fearful aspect, as it charged and charged again through the seething masses of the enemy, leaving a trail of writhing or stiffening victims in its course.

At last its own fate came. In a heedless moment it stumbled over a wounded foe, whose jaws at once closed, and closed for ever, on its leg. Reeling backwards, it fell into the very midst of three or four more, and hope of escape was gone for ever. They threw themselves on it like demons, and though it rolled on its back amputating and decapitating until limbs and heads and headless trunks strewed the ground, all the fury of despair was of no avail against the numbers that continued to heap themselves on it. At length its struggles grew feebler and feebler, its ponderous jaws opened and shut slowly, like some animate rat-trap sighing for rats, and its life ebbed away. The scene was Homeric, and I felt like breathless Jupiter watching Hector on his fatal day, when he felt the movings of pity, yet let fate take its course. This was an epitome of the whole struggle. It must have raged all night, but neither side got victory. In the morning each was in quiet possession of its own ground, and the fruits of the battle were many hundred corpses and a moral.

Solomon has advised us, or most of us, to go to the ant and consider her ways, and it is good to follow his advice. Her ways repay consideration. But it is of vital importance that we go to the right sort of ant. What a lesson, for instance, in malice and all uncharitableness would one learn who went to the red ant which infests the corrinda-bushes on Matheran and Khandalla, or on the slopes of Elephanta Island! Malice, hate, fury and fierceness, wrath and rancour, acerbity, and, in fact, every feeling which is out of harmony with 'sweet reason', seems to have been boiled down, and its quintessence extracted to compose the blood which courses angrily through the hot veins of this creature. As you pant up the red-dusty path, the red ant hears you afar off and hurries along the outermost branch, to the very point of the very longest leaf, and there stands on tiptoe, dancing with impatience to bury its jaws in your flesh. And what a knowledge it has of our geography! What an instinct for detecting tender places!

Industry is not to be learned from these. I believe they lead idle lives and live on the milk of their flocks and herds. In the month of May, when the corrinda-bush is in fruit, I have often noticed with pain that the choicest berries were in possession of a garrison of red ants, which had enclosed them in a sort of chamber by drawing the surrounding leaves together and joining them with some spider's-web fabric which they spin. This is not for the sake of the fruit. They are not frugivorous. It is for the sake of the downy white aphides, or plant-lice, on the fruit. These aphides yield a sort of nectar, which is as delicious to an ant as camel's milk is to an Arab. But other ants are content to milk the unresisting little cattle whenever they find them; the red ants domesticate them.

The ant to which Solomon sent sluggards was plainly the agricultural ant which lives in the fields. A space of ground round the mouth of its hole, about as wide as the hat of a padre whose views are beginning to get ritualistic, is always cleared, like a threshing-floor, and covered thick with the husks and chaff of the grain stored inside. These holes are the gateways of great cities, and from them broad well-beaten roads lead away in all directions to other distant cities. Late and early these roads are thronged with crowds of busy ants. As I sit and watch them on a sunny morning, the primitive *ryot* stops shrieking at his perverse *byle*s, and for a moment puzzles his foggy brain to guess what I am doing. He believes I am on the scent of hid treasure, but his more intelligent neighbour says I am simply illustrating the inscrutable ways of the sahib.

I confess I lean towards Sir John Lubbock's view that ants are gifted with reason like ourselves. There is no objection to explaining the wonderful things they do by instinct, but only a new meaning will have to be invented for the word. The instinct which a weaver-bird shows in building its wonderful nest belongs plainly to a different genus from the quality which enables ants to 'vote, keep drilled armies, hold slaves, and dispute about religion', as Mark Twain says they do, or even to talk. They certainly do talk about as freely as we do. I once killed a centipede, and very soon a foraging ant found it. He, or rather she, surveyed it carefully, estimated the horse-power requisite to move it, and then started off homewards. Meeting another ant, she stopped it and said something which, for want of a microphone, I did not hear, and hurried on. The second ant made straight for the centipede and found it without any trouble. Now nothing can be plainer than that the first ant told the second where to go. 'Glorious windfall! Dead leviathan about two miles from here. Keep straight on till you come to a three-cornered pebble, then turn to the left and you will come upon three grains of sand and a straw. Climb the straw and you will see it. It is big enough to be seen a mile away.' Well, the second ant, when it had found the centipede, did not hurry home. It just sat down and waited till the first one returned, with a vast gang of labourers; then each seized a leg of the centipede, and soon the stupendous mass was moving along merrily.

But not only has each species of ant a language in which it can talk to other ants of the same species, but each nest, or clan, has clearly its own brogue; for an ant knows at once whether another ant belongs to its own nest or not. The ants of one nest murder those of another: it is a point of honour with them.

There is no mode of life that men have tried which one race of ants or another is not pursuing today. Besides those which are agriculturists or herdsmen, some keep slaves to do everything for them, some live

---

Ryot (*raiat*) tenant farmer   *byle* (*bail*) ox

by hunting or plunder, while others quarter themselves on us and subsist by confounding *meum* and *tuum*. These last, of course, concern us most. About Bombay there are two kinds of them, one black and the other brown. They are both small, and most people confound them, but in nature they are antipodal. There is not any figure or simile which can even dimly shadow forth the extent of their oppositeness. Chalk and cheese are the same article by comparison. That ignorance should prevail on this point, even amongst persons who have undertaken the responsibility of housekeeping, is distressing, for it borders on criminality. In a healthier state of public opinion a young lady would not be considered 'eligible' who could not converse freely on the difference between the black and the brown ant. That difference in its essence is this, that the one is tolerable and the other intolerable. If one must go into more detail, the brown ant is thickset, heavy, slow and phlegmatic. It will eat, more or less, everything in the house except, perhaps, kerosene oil. It will gnaw a cold leg of mutton, carry excavations into the heart of a loaf of bread, dig a tunnel through the cork of an olive-oil bottle, for the sake of getting drowned in the oil, and organize a regular establishment for the work of carrying off the seed in the canary's cage. And, once in a thing, it cannot be got out. Add to this that it smells unsavoury and tastes nasty, and you have the brown ant. The black ant is slender, nimble and sprightly. Its chief business in the house is to remove dead cockroaches, crickets etc., and where I am there is generally a plethora of dead cockroaches, crickets etc. All day foragers scour the house in search of these. They do tamper with sugar sometimes, and, in fact, show a leaning towards sweets in general; but they do not spoil what they cannot eat. They do not stick, as a rule, in the jelly, nor drown themselves in the ginger syrup. Lastly, there is a feud between them and the brown ants, and the two will scarcely live in the same house. Clearly, then, it is sound policy to make an ally of the black and discourage the brown. The latter is not an easy task, but I can recommend dropping kerosene oil into their holes.

The large black ant, already mentioned, is more or less a house ant also. I do not like it. The way it cocks its tail over its head is offensive, and it has a cantankerous temper. Then its officiousness and consequential airs are simply insufferable. It is perpetually quarrelling with a straw or getting insulted by a feather.

Meum mine   *tuum* yours

Of all the various species of these wonderful little beings, there is not one, I think, that impresses you more than the hunting ant. It is, unfortunately, not a house ant. It just invades the house at times, does its short sharp work, and is gone again. In these expeditions they always march in a column, three abreast, with rapid steps and terrible earnestness of purpose. Not one wanders or lags behind. Sugar entices them not; stores have no attractions for them. Straight as General Roberts they make for some ancient trunk in whose chinks and crannies the outlawed cockroach and overgrown cricket have long skulked secure from my avenging slipper. Now their hour is come. With the rapidity of perfect system a guard is stationed at each hole and crevice, and then the main body of ants pours itself into the box. Then begins a panic. The cockroach, wild with terror, rushes head-long to the nearest outlet, and is collared by the guards and stung to death almost before it has time to realize the situation. The frantic crickets break into coruscations of agility which would enable one who has never seen an aurora borealis to realize it. But all is vain. Within a quarter of an hour the ants are marching out as they marched in, three abreast, with rapid steps; but now, with drooping limbs and trailing antennae, cockroach and cricket, cricket and cockroach, follow the long column in funeral procession.

The Frogs (December)

In the columns of the *Times of India* I have had a public invitation from 'Sarus' to descant upon Frogs. I had thought to pass the vile batrachians by, for I love them not. Besides, now is not their time. The hot sun has been boiling down the tanks until the infusion of frog is getting thick, and the water-snake grows fat on much to eat and little to do. So the bass-voiced patriarchs of the tribe have dispersed to many secluded water-holes, or perhaps have buried themselves in the mud, and even the nimble small fry, skimming with many a hop, skip, and jump along the surface of the water, have much ado to save their lives from the fierce fish and the remorseless *dhaman* below, not to speak of the gluttonous heron above. Were the rain at this moment dripping from the roof and gushing from the water-spout, and a concert of a hundred bassoons from the flooded paddy-field sounding in my ears, I could write on frogs.

Dhaman kind of water snake

I believe the observations of 'Sarus' are vitiated by the common mistake of confounding things which differ *toto coelo* from each other. To take, for instance, the frog which he found on the top of a door, it is obvious that everything turns on the question: Had it, or had it not, little round pellets on the points of its toes? If it had not, then it ought to have been bottled in spirits and sent to the able and energetic secretary of some learned society, for a common frog which can climb to the top of a door ought to have an essay written on it. If it had, then it was only a tree-frog, a species which was rather a favourite with me until one evening last year. There were several of them about my house and their gymnastics won my admiration. From a yard away they would fling themselves at a bedpost or a window-pane, and stick like a dab of mud by virtue of these suckers on their toes. They would perch pleasantly on the edge of the water *cooja*, or on the rim of a tumbler. They seemed to gain little by all their performances, for their appearance was always famine-stricken and angular, and their colour, without being very definable, suggested the sere and yellow leaf. They slept all day, sticking like postage stamps to some window, and at evening went abroad in search of food, leaping from one perilous position to another about the railings of the verandah.

On the particular evening above-mentioned I was sitting in the garden, trying to finish a very interesting chapter in a book before it got too dark to read—at least, I imagine that was my occupation, but my memory about that period is almost a blank. Within a few feet of me there was a projecting sunshade, and on it clung an enterprising tree-frog. To him my head loomed like some forest-clad mountain against the grey sky, and he guessed there might be game up there. So he wound up his leaping springs, took good aim, allowed for the wind, and fired. I do not know exactly where he aimed, but he hit just behind my right ear, and, of course, stuck. Now, I hold that half the art of telling a story, as of preaching a sermon, lies in knowing when to stop, so I will stop; suffice to say, that since that evening I have admitted no exception to the general feeling of utter aversion with which I regard the whole race of frogs.

To proceed to the so-called frog, which comes into the house and out-generals 'Sarus' on his attempts to evict, I know it well. It is not a frog at all, but a toad. The difference between the two is precisely

---

Cooja earthenware water-container

the difference which there was in Mark Twain's jumping frog before and after the shot was administered to it. Touch a frog ever so tenderly with a stick from behind, and it goes off as if it were sitting on gunpowder, and your stick were a lighted fuse. The stolid toad, on the other hand, meets every hint and every suggestion with a simple *vis inertiae*, and an unwavering perversity and 'contrariness', which must triumph in the end. Now, when a man has made up his mind beforehand what his final opinion is to be, it is waste of time to dispute it with him; therefore I always clinch the argument at once with my toad. I bully him until he feels thoroughly affronted, and refuses to budge another inch, blowing himself up like an air-pillow, and snorting feebly by way of protest. Then I induce the point of a springy cane under him, and simply shoot him out of the door. He takes it very ill, but I cannot help that. It would be mistaken kindness to let him delude himself with the notion that he is going to get what he wants in the house. I know exactly what it is. As the cold, dry, easterly winds begin to shrivel and crack his parched hide, it crosses his foggy brain in some dim way that a house must contain a lot of cool damp holes and corners, into one of which he may wedge himself, and pass the dry months in a state of torpor, conserving his vital juices till next monsoon. This is a proposal, of course, which cannot be entertained. He is all very well flattened out under a flower-pot or between the stones of a fernery; but it is preposterous to suppose that he can be allowed to take up his winter quarters inside the house, and I think the most considerate course is to impress this on him before he has crossed the very narrow line that separates his normal state from actual unconsciousness.

For I have kindly feelings towards the toad; the density of his stupidity, and his placid contentment, make ill-will towards him impossible. Low-bred he is, but more than half the world must always be low-bred; there is no crime in that. No sane man pretends to despise another merely because he is low-bred; unless indeed he feels that his own high breeding stands in need of a contrast to make it visible. Ostentatious vulgarity is a very different thing, and it is this that makes the frog an offence to me. He is for making a noise in the world. He will thrust his gross entity on your notice. If the rain, which dampens everything else, only cheers the spirits of frogs, I have no objections; let them be happy. But why must they, with their riotous cacophony, proclaim the fact to the world, after the manner of ' 'Arry'?

Further, I have physiognomical objections to frogs. The aspect of them is an outrage. Every line of their gape-mouthed shallow-pated visages bears witness to general debasement, and an inordinate love of victuals. The little leopard-spotted water-frog is more tolerable; but I am speaking of the gross overgrown bull-frog. After months of bleaching—while it lay torpid, I suppose, in the ground—it comes out to greet the monsoon all of one uniform gamboge yellow, and riots in the daytime. Then, when lusty health has restored it to a dark green hue, with a gaudy yellow line running down its backbone, it leads an amphibious life, lurking among the rushes on the margin of some pool, and at the sound of your footstep taking a header into the water, with its legs, like the tail of a comet, behind it; or, perchance, having tumbled, during some ill-fated spree, into a deep well, it expiates the crime of its appearance by a long life of solitary confinement, with no hope of release. The livelong day it is doomed to float at the surface of the water, vacantly gazing at heaven, with supplicating palms outstretched and fat thighs helplessly pendulous in the clear liquid; but sudden death is oftener the frog's fate than imprisonment. Every one will call to mind the case of the young rip whose amorous career was cut short by the lily-white duck that gobbled him up; and herons are worse than ducks, for they do not wait till he goes a-wooing, but stalk into his haunts, and from the far-darting serpent neck and scissor beak of a heron escape is hard. Then the marsh harrier pounces down among the rushes on the croaking veteran who had outlived these perils, and bears him away in its talons. But the arch-enemy is the dhaman, or water-snake, and it is more cruel than the rest, for it takes an hour or two to swallow its victim. It is impossible to conceive a fate of more unmitigated horror than that of a frog being sucked down by a snake, its foot already undergoing digestion, its legs stretching all the way down the enemy's slimy throat, and its body slowly but surely following. Happily frogs cannot have much imagination, yet they must realize the situation to some extent, for they give expression to the anguish of their souls every few minutes in a wall so unspeakably woeful, that it would melt the hardest heart. It has often melted mine to such an extent that I have gone out with my stick to slay the snake, and release the frog. Once I saw the tables turned. I was watching a wily snake about two feet long gliding down into a tank, when a gigantic frog hopped up and swallowed its head. The snake protested with frantic wriggles, but the frog continued swallowing it down—an inch or two at each gulp—until half the snake was gone. By this time the other half became so violent that the frog could scarcely keep its feet, so for greater security it turned and plunged into its own element, and I saw it no more. Even this was beaten in audacity by a frog from whose stomach I, David-like, redeemed the whole leg of a live chicken. The rest of the chicken was still outside, remonstrating clamorously.

Of frogs for the table I have said nothing, having no experience, for I look upon it as cannibalism to eat them until the question has been finally decided whether we are more immediately descended from them or monkeys.

The Crows

What is there that can be said about them? Have they not sufficiently cast a shadow on our lives, left their black mark on our pleasantest memories, yea, even their scars on our dispositions and tempers? Yet it is impossible to pass them over. I can call up no vision of Indian life without crows. Fancy refuses to conjure up the little bungalow at Dustypore in a happy state of crowlessness. And if the mind wanders away to other times and distant scenes, the crow pursues it. It is sitting impudently in the hotel window, it is walking without leave in at the open door of the traveller's bungalow, it is promenading in front of the tent, under the mango *tope*. Only when in thought we go back to happy rambles away from the hum of men,

Where things that own not man's dominion dwell,
and mortal foot hath ne'er or rarely been,

is the horrid phantom absent. On the breezy hill-top, with its scented grass, its ferns and wild flowers, down in the solemn ravine, where the 'whistling schoolboy' tunes its mellow throat and the clucking spur-fowl starts away among the rustling leaves, you meet no crow. The air is too pure and the calmness too sweet. The crow is a fungus of city life, a corollary to man and sin. It flourishes in the atmosphere of great municipalities, and is not wanting in the odorous precincts of the obscure village innocent of all conservancy.

Many of our frontier tribes have unpleasant traits of character, and in some the catalogue of vices is long and the redeeming virtues are

---

Tope grove

few. But the crow differs from them all in that it is utterly abandoned. I have never been able to discover any shred of grace about a crow. And what aggravates this state of things is the imposture of its outward appearance. It affects to be respectable and entirely ignores public opinion, dresses like a gentleman, carries itself jauntily, and examines everything with one eye in a way which will certainly bring on an eyeglass in time, if there is a scrap of truth in the development theory. But for this defiance of shame one might feel disposed to make allowances for the unhappy influences of its life; for, in truth, it would be strange if a crow developed an amiable character. Even a consistent career of crime must be less demoralizing than the aimless vagabondage by which it maintains itself. It begins the day by watching the verandah where you take your *chota hazree*, in hope to steal the toast. When that hope is disappointed it wings its way to the bazaar, where it contends with another crow for the remains of a dead bandicoot flattened by a passing cartwheel. Then, recollecting that the breakfast hour is near, it hurries back, not to lose its chance of an eggshell or a fishbone. On the way it notices a new-fledged sparrow trying its feeble wings, and, pouncing down ruthlessly, it carries the helpless little sinner away to a convenient bough, where it sits and pulls it to pieces and affects not to hear the pitiful screams of the heartbroken parents. Later on it is watching a little stream of water by the roadside and plucking out small fishes as they pass, or it is vexing a frog in a paddy field, or it has spied a swarm of flying ants and is sitting down with a mixed company to supper. For another instance, take the following, which I myself witnessed, and say if anybody could have a hand in such a transaction and preserve his self-respect. A large garden lizard had wandered unwisely far from its tree, when two crows observed it and saw their advantage. They alighted at once and introduced themselves, like a couple of card-sharpers. Then the lizard also took in the situation, and, wheeling about, made for the nearest trees. 'Not so fast,' quoth one of the crows, and with three sidelong hops, caught the tip of its tail and pulled it back again. Then the lizard reddened to the ears with offended dignity, and swelling like the frog in the fable, squared up for the fight; for lizards are no cowards. But the crows had not the least intention of fighting. They remained as cool as cucumbers and merely took up positions on opposite sides of the lizard. The

Chota hazree early morning tea and fruit

advantage of this formation was that, if it presented its front to the one, it had to present its tail to the other, and so, as often as it charged, it was quietly replaced on the spot from which it started. Now, to be continually making valiant rushes forward and continually getting pulled back by your tail must be very discouraging, and after half an hour or so the lizard was evidently quite sick of the situation. But as its spirits sank the crows' spirits rose. Their familiarities grew more and more gross, they pulled it about, poked it in the ribs, cawed in its very face and finally turned it over on its back, with its white breast towards the sky, and were preparing to carve it, when suddenly the squirrel gave a shrill warning, a panic seized the hens, and the two miscreants just had time to dart aside, one this way and one that, as a kite, with whirlwind swoop, dashed between them and bore away the lizard in its talons. They stared after it with a gape of utter nonplussation.

I do not know about the Afghans, but a policy of masterly inactivity will not do for the crows. Their peculations and insolence always extend to the limits of your toleration, and they keep themselves acquainted with those limits by experiment. I go in for keeping up my prestige with them. I shoot a crow once a month or so and hang it up *in terrorem*. This has such an excellent effect that no crow ever sits on my window and gives three guttural caws in the caverns of its throat, with intent to insult, as they do at other people's houses; nor are their evening convocations holden on my roof.

In April and May crows make nests of sticks and line them with *coir*, or horsehair extracted from a mattress, or even with soda water wire stolen from the butler's little hoard! In these they bring up three or four callow criminals in their own image. I make all such proceedings penal about my premises, for the claims of a hungry family will drive crows to even more reckless wickedness than their own inbred depravity. They will appropriate hens' eggs, murder nestling pigeons, attempt the life of the canary, and every now and then startle you with some entirely new and unthinkable felony.

Most young things in nature are engaging. We grow more unlovely as we grow older. What is prettier than a downy chicken, a precocious kid, a young mouse not an inch long, or that little woolly image of comfort, an infant rabbit, when it first shows its round face at the door of the nursery? But new-fledged crows are a staring exception to the rule. They are graceless crudities, with glazed eyes and raw red throats, which they show you about three times a minute, when they open their mouths to emit an inane caw. They should be put to death offhand.

All the above remarks refer of course to the grey-necked crow. To make them applicable to the large black crow, they must be discounted ten to fifteen per cent. There is some sturdiness of character in the black crow; it is a downright, above-board blackguard and my feelings towards it have some semblance of respect.

A Dog's Day

A dusky queen from Aberdeen,
she quitted home and clan
to cheer her lone adorer
in cheerless Hindostan;
Fate drew the contract for her,
and thus the wording ran:

'A hapless two-legged creature
pursues his fortune there;
his days are void of feature,
his life is full of care.

For men whose hearts are double,
whose honeyed words are lies,
in sickness and in trouble
his weary task he plies.

He needs the conversation
to dogs and masters dear,
the eye of adulation,
the tail that says, 'I'm here.'

Part of the white man's burden
the white man's dog will bear,
without the hope of guerdon
that falls to human share.

No matter what his choice is,
his whimsies she'll respect,
rejoice when he rejoices,
if he reflects, reflect.

For twelve long years untiring,
till age lays waste her might,
she'll serve him, his admiring
playmate and satellite.

And ere the man she's tended
hies on his homeward quest,
her dole of life expended,
she'll lay her down and rest.'

O eyes so bright and pleading,
O mien sedate and high,
so soon to darkness speeding—
how deep your mystery!
Why wring this heart with your black art,
then lay you down and die?

A.H. Vernède, ICS, 1920

7
Translations

The following excerpts are from A.G. Shirreff, *Tales of the Sarai*. The Introduction and Conclusion are by Shirreff. The tales are translations by him.

Introduction

    Some two miles west of Kasia
there rises o'er the level ground
a lofty many-acred mound,
which once was Kusinagara.
Beneath it buried lies the frame
of court and cell and cloister built
ere London was a local name
or Paris rose above her silt.
    India has had in every age
many a place of pilgrimage.
From where the tower of Jagannath
sees sunrise o'er the eastern wave
to where the western waters lave
the pillaged portals of Somnath,
from Haramukh's eternal snows
to where the sundering channel flows
round coral-reefed Rameshwaram,
to many an immemorial fane,
by road and river, hill and plain,
innumerable pilgrims come.
    But east or west or south or north,
no place of pilgrimage there is
whose contemplation should call forth
more holy memories than this.

For him whose tomb is in this place,
after five times five hundred years,
a third part of the human race
above all human kind reveres.

    Yet for a thousand years or more
the tomb lay hidden and forgot,
till late-deciphered eastern lore
led western science to the spot.
And now beneath the *stupa* stands,
upon the level ground hard by,
shaded by trees, a small *sarai*,
where pilgrims meet from many lands,
and in its shrine when they alight
are wont to make an offering
of jewelled sandals richly dight,
in token of their travelling.

    Here, on a showery winter day,
soon after noon, four men had found
a shelter in the covered way
that closed the little courtyard round.

    The first your glance had rested on,
if you had seen that group, was one
whose shaven head and saffron dress
proclaimed his calling's sacredness;
an old man in whose gentle mien
wisdom and kindliness were seen.
The Abbot this, if so we may
term one whose rule no monks obey;
for though at times some novices
sat at his feet, or three or four—
there scarce was living room for more—
at present there were none of these.
The duty which concerned him most
and most delighted, was to be
a ready guide, a courteous host
to travellers of each degree.

   Stupa a large round earthen mound marking a Buddhist shrine and centre of pilgrimage   *sarai* traveller's hostelry, with stables

One Pilgrim here you had espied,
clothed in a long sad-coloured coat,
wide-sleeved and narrow at the throat,
upon the Abbot's left-hand side.
His speech and tone to Indian ears
would scarce reveal the Chinaman,
for he had travelled now for years
the length and breadth of Hindostan,
visiting every sacred place,
and every undistinguished mound,
where of the Master any trace
or of the early faith is found.

    Facing him, on the Abbot's right,
clad in a flowing garb of white,
a yeoman sat, of *thakur* caste,
who lived about a mile away,
and often with the Abbot passed
part of a not too busy day;
a Surajbansi, who could trace
the first beginnings of his race
further than Hapsburg or than Guelf
beyond the legendary kings
of whom the *Mahabharat* sings,
right backward to the Sun himself.

    With them there was in company
a Musulman whom his *chapras*
(the badge, upon his belt, of brass,)
declared to be an Orderly.
He had come on from Kasia
with tents and carts provided there
his master's lodging to prepare,
who from the rail at Deoria
had five and twenty miles to drive,
and after nightfall would arrive.
His master was a learned man,
a travelling American,
who in some University

Thakur The warrior caste closely associated with the nobles of the Rajput race

far in the doorways of the west
had studied Sanskrit, and professed
comparative philology.
But to the servant I return:
the master is not my concern.

    He too could boast a kingly line
made western pedigrees seem tame,
and of his ancestor could claim
Sultan Sikandar Zu'l Quarnain,
Sikandar, King of Macedon,
Sultan Filipus' mighty son,
who, aided by the sorcery
of Aflatun, his chief wazir,
subdued all kingdoms far and near,
and even took tribute from the sea.
For he was of the khetwar clan,
from the Punjab, up Pindi way,
and his employment first began
upon a tour to Taxila,
with which in view he had been lent
to the Professor for a term,
but soon induced him to confirm
his services as permanent.
A merry, lively rogue was he,
welcome in any company,
but most when nights are long and cold,
and round the camp fire tales are told;
for many a story he could tell
of magic carpets, lamps and rings,
Djinns and Afrits and ancient kings,
and by long practice told them well.

    These were the four. The Orderly
was speaking to the other three.

    'Truly,' he said, 'this little place
though so sequestered, has no dearth
of visitors of every race
and from the ends of all the earth.
This night there will be two at least,
my master from the furthest west,

and you, sir, from the furthest east,'—
(this to the Pilgrim was addressed,
whom he had just been questioning
about the Peiho and Peking,
and of the fortunes of the war
to which his brother had been sent
some twelve or thirteen years before
in a Punjabi regiment).
'Here must the man whose lot is cast
learn many a tale of many a land;
here too is ever close at hand
our country's immemorial past.
Far have I been, but never yet
have seen so apt a spot to whet
a story-telling appetite;
a circumstance that suits me well,
for tales to hear and tales to tell
has ever been my chief delight.
How say you then, Sirs? I suggest,
since here it seems that we must stay
for the remainder of the day,
that we can spend our leisure best,
and respite best from tedium earn,
by telling each a tale in turn.'

  With readiness the rest approved
of the proposal he had moved,
though some demur the Thakur made.
He was, or said he was, afraid
his stay-at-home and rustic wit
for such a task was little fit.
And to the Abbot he appealed,
who, straight betraying him, revealed
how no-one in that country-side
gentle or simple, had more store
of song and legendary lore;
a charge which he in vain denied.

  The Orderly, it was agreed,
should, as the author of the plan,
in its performance take the lead;
who, straight consenting, thus began—

The Orderly's Tale: The Counter of Bubbles

Akbar the Emperor held his Court
on the Jumna's bank in his new-built fort.

Hundreds of applicants thronged the hall;
he heard them patiently one and all.

There stepped at last o'er the sandstone flags
a hungry rascal in beggarly rags,

who cried, 'Your Majesty, pray protect
a wretch whose fortunes are utterly wrecked.

This tattered garb my condition speaks;
I haven't had a square meal for weeks.

My father left me a fine estate,
which I soon proceeded to dissipate.

I have spent my inheritance, sold my land,
run through my credit,—and here I stand

in the hope and trust that your Majesty can
find me a job for a gentleman.'

'A job?' said the Emperor. 'Let me see.
What may your qualifications be?'

'My qualifications? None, alas,
but those that have brought me to such a pass.

Feasting and flirting and roistering,
drinking and dicing and cock-fighting

are honest accomplishments no doubt,
but nothing exactly to brag about.

I have no skill to cipher or write;
I have no stomach to serve or to fight.

I would blush to beg and I cannot dig,
as for setting up shop, I had rather prig.

I'm fit for nothing at all, in short,
except for some sinecure at the Court.'

The Emperor Akbar turned to where
Birbal the jester stood by his chair.

He whispered apart with him awhile,
then turned to the applicant with a smile.

'I have found,' he said, 'a vacancy
suited to your capacity.

No salary is attached, it's true;
but there's plenty of honour and little to do.

You can count the bubbles that float down stream
without impairing your self-esteem.

And if the post suits you, I will instal
you as Bubble Counter Imperial.'

'*Wah wah! Shabash!*' cried the Courtiers then.
'Truly our King is a king of men.

See how he knows the way to deal
with such a ridiculous ne'erdoweel.

The knave would do well to lay to heart
the lesson these jesting words impart;

that the pleasures his wealth has been wasted on
are lighter than bubbles and sooner gone;

and nothing can be accomplished by
sitting to watch life's stream run by.'

But the applicant, not abashed a whit,
said, 'I like the billet. Pray give me it.

And to make the appointment right and tight,
let me have a *Sanad*, declaring my right.'

Said the Emperor laughing, 'Where's the harm?
Give him a Sanad as long as his arm.'

So said, so done, and the rascal went
out of the audience hall content.

. . . .

Two years passed, and in Agra Fort
Akbar again was holding his Court,

Wah exclamation of praise   *shabash* bravo   *sanad* certificate, charter

when with vast disturbance and dust and din
a splendid procession swaggered in.

At its head was a proud magnifico
blazing with jewels from top to toe.

Said Akbar, 'Who is this nobleman?
I can't recollect him, try as I can.'

The courtiers were all alike non-plussed,
when through the crowd a retainer thrust,

proclaiming, 'Way! Make way for the late
Counter of Bubbles to Akbar the Great!'

With a rustle of silks and a blaze of gold
up to the dais the visitor strolled,

and said, 'Be pleased to resume, my Prince,
the Sanad you gave me two years since.

Greed was never a vice of mine;
that is the reason why I resign.

Now I count my fortune by Lakhs and Crores,
and I owe it all to this gift of yours.'

'Tell me,' said Akbar, 'how was it done?
For surely the post was an unpaid one.'

'I started,' he said, 'with the obvious fact—
to be counted, a bubble must be intact;

but up stream or down stream, fast or slow,
without breaking bubbles, no boat can go.

And to hinder the Emperor's work and mine
deserves, at the least, a substantial fine.

My charge was an asrafi per bubble.
If anyone argued, I fined him double.

For the stoutest objector had to yield
when he saw the deed you had signed and sealed.

Lakh   one hundred thousand   crore   ten million

There is many a boat goes up and down
where the Jumna passes by Agra town;

and I found my duties a paying game;
there are others, no doubt, who do much the same;

for your subjects pay more in revenue
than ever reached Todar Mal or you.'

The King said nought, but his brow grew black;
and his courtiers whispered behind his back,

'He has dished himself by his own device,
that comes of following Birbal's advice.'

The Thakur's Tale: The Hypocrite

The Gupta king sent his envoys forth
to fetch him a *Pandit* from the North;

from Pataliputra, his abode,
up to Cashmere by the Grand Trunk Road.

(The Grand Trunk Road was not built, I know,
not by twelve hundred years or so;

but there must have been, if not the same,
a similar road with a different name,

else how could the Guptas have spread their power
from the Hugli's mouth to beyond Peshawar?)

Up to Cashmere the envoys went,
found the sage for whom they were sent;

brought him down to Behar again,
over the passes and down the plain;

Jumna river in north India   Todar Mal Akbar's very able revenue minister   Gupta dynasty ruling India between *c.* AD 314–540   *pandit* a learned man, typically a brah-min   Pataliputra capital of the Maurya and the Gupta dynasties   Hugli (Hoogly) a river in the western branch of the Ganges delta, which flows into the Bay of Bengal   Behar a province of British India, bordered by UP in the west, Bengal in the east, MP in the south, and Nepal in the north

but a private audience first they sought,
before to the presence he was brought.

'We are sorry,' they said, 'to mention it,
but your Pandit, Sire, is a hypocrite.

When we found him first, we are pained to say,
he was sitting to eat on the public way;

he had not bathed as a Brahman should,
and Heaven only knows who had cooked his food;

he had not stripped to a single sheet,
but was clothed in furs and had boots on his feet.

We do not complain so much of this,
for all Brahmans up there are alike remiss;

but, once in the plains, he displayed his zeal
by stripping to bathe before every meal;

and still his propriety has increased
each stage we have travelled South by East.

He has played at observances like a game,
aping the custom wherever he came,

till here, where all Brahmans the rule observe,
not by a hair's breadth does he swerve;

he cooks his food at a stove of his own;
strips to a loincloth of silk alone;

feeds in his own particular square;
never eats food that is cooked elsewhere;

makes the oblation to *Thakur Ji*;
in short, is as strict as strict can be.'

'Well,' said the king, 'I take it, you
kept to the strictest rules all through.'

'To be sure we did,' their spokesman says,
'in spite of a hundred hindrances.

Thakur ji God

Some of us caught catarrhs and chills
from stripping to feed on the frozen hills;

and all of us fasted two days and more,
for want of cow-dung to plaster the floor.'

The king's reply was direct and terse;
he simply quoted a Sanskrit verse,

which means translated, 'The more fools you
not to do in Rome as the Romans do.'

The Orderly's Second Story: The Faith Cure

Dragging on a pair of crutches his emaciated frame,
to the *Pir* at Pipra ferry Ahmaq the *julaha* came.

'Holy Saint,' he cried, 'have pity, and exert thy power to save
one whom magic arts are hurrying prematurely to the grave.

As you know, the *Tharu* women all possess the evil eye,
and its pitiable victims are most often such as I.

For these witches have a weakness for a handsome bachelor
such as I was only lately,—such as I shall be no more.

There's a fatal fascination in the Tharu woman's glance;
if she once has overlooked you, you don't stand an earthly chance.

There's a buxom Tharu widow who sells fish in Tulsipore;
she has cast the glamour on me, I am absolutely sure.

See, my limbs are all a-tremble, and my skin is ashen grey;
all my strength is turned to water; I am withering away.

These are symptoms of the Lohna, which invariably ends
in excruciating torments, as I hear from all my friends.

Soon my liver, being shrivelled to the bigness of a pea,
will be riven from my midriff, which will be the end of me.

Be the end? I wish it were, though. After death I shall be still
everlastingly the victim of that wicked woman's will.

Pir Muslim saint   *julaha* weaver   *Tharu* gypsy tribe living in the Terai
and Bhabar forests at the Himalayan foothills

I shall flitter at the midnight through the glimmering forest glades,
speeding on her gruesome errands in a troop of gibbering shades.

It is you alone can save me from this miserable fate.
Holy father, have compassion; help me ere it be too late.'

'Yes,' the sage said, 'I can heal you, if my bidding you obey.
Tell me first, how came you hither, and on which side of the way?'

Ahmaq wondered at the question. 'I arrived here,' he replied,
with the aid of these two crutches, keeping to the left-hand side.'

'This then,' said the sage, 'will cure you. Go straight back to your
    abode,
walking only on the *patri* on the right side of the road.'

'Is that all?' said the Julaha. 'That is all,' replied the Pir.
'One thing more, though—those two crutches: you had better
    leave them here.'

Ahmaq did so and departed, crawling painfully and slow,
but he felt a vast improvement after half a mile or so.

By the time he passed Turkaulia health was glowing in his cheeks;
and he reached his home an-hungered as he had not been for weeks.

with a day or two's high feeding when his tone was quite restored,
back he journeyed with a present of the best he could afford.

'Twas a web of his own weaving worked upon a special plan,
all the colours of the rainbow rioting in every span.

When the gift had been accepted, Ahmaq said, 'If there is nought
unbefitting in the question, tell me how the cure was wrought.'

Said the sage, ''Tis very simple. As you journey back to-day,
you will notice there are *nim*-trees on the right side of the way.'

That the nim has healing virtue you no doubt already know.
To that virtue's efficacy your recovery you owe.'

Back went Ahmaq, and thenceforward night and day his constant
    theme
was the wisdom of the hermit, and the virtue of the nim.

Patri footpath

But the Holy Man's disciple, as the patient went away,
said, 'With your permission, Master, there is something I would say.

If he journeyed on the left side when from Tulsipore he came,
and returned upon the right side, either way it was the same.

Either way it was the same, and there's no reason on this showing,
why the nim-trees should have cured him not in coming but in going.'

Then the sage of Pipra ferry answered smiling,—and I think
that he must have winked, if ever holy sages deign to wink—

'Yes, I nearly made a blunder; but the risk was very slight.
Was there ever a julaha knew his left hand from his right?

For the nim-tree's power to cure him—you may doubt it if
    you will;
it is every whit as true as Tharu magic's power to kill.'

Conclusion

The time of sunset now drew near
and long ere this the skies were clear;
so, when the tale had reached its close,
after some desultory talk,
the Thakur took his leave and rose
homeward across his fields to walk;
the Orderly to supervise
the pitching of his master's tent,
and, fingering their rosaries,
the others to the chapel went.
The Thakur, as he strode along,
lifted his voice in a last song.—

What is past is past; forget it. Only think of what's to come.
Let what effort still may compass be of all your cares the sum;
be of all your cares the sum what may be in the end achieved.
So man's malice shall not mock you, neither shall your heart be
    grieved.
Thus says Girdhar, prince of poets, pin your hopes and purpose
    fast
to the pleasures of the future, knowing what is past is past.

A.G. Shirreff, *Tales of the Sarai*, 1918

Translations from Martial

IX 98

The vintage was not all in vain:
Coranus found the downpour gain,
and bottled eighty quarts of rain.

VIII 12

'What, marry an heiress, and pocket my pride?
Why, she would be bridegroom and I should be bride!
Though mine be the wit and the rank and the riches,
her chance is still even of wearing the breeches.'

T.F. Bignold, ICS, *Leviora*, 1863–88

Translation from the Hindi

The son of a *Kaiasth*
    will always be biassed:
or, if he's not biassed,
        his father's no Kaiasth.

A.G. Shirreff, ICS,
The Dilettante, 1918

Resma and the Kotwal

Resma was her daddy's darling, and her daddy for a treat
often gave her, as a favour, pounds and pounds of cloves to eat.

*Refrain: Resma darling, Resma darling, Resma darling, by and by
    somebody will be your husband. How I wish it might be I.*

In her silk-embroidered bodice and her petticoat of blue
she was neat and she was sweet and most attractive to the view.

Resma went one day to market in the dress described above,
and the *Kotwal* fell a total victim to the power of love.

'O my golden girl!' he babbled, 'tell me from what perfect mould
thou departest and what artist formed thee of the purest gold?'

Kaiasth a caste group in northern and north-eastern India, prominent in business

'O you silly old policeman, I should like to burn your beard,
with your golden girls and moulding. Stuff and nonsense!'
  Resma jeered.

'One half of me by my daddy, if you want to know, was given
and the other by my mother; but my beauty comes from heaven.'

*Resma darling, Resma darling, Resma darling, by and by
somebody will be your husband. How I wish it might be I.*

A.G. Shirreff, *Hindi Folk Songs*, 1936

Nursery Rhymes

If all the world were apple-pie,
and all the seas were ink,
and all the trees were bread and
    cheese,
what should we do for drink?

*Darya shor siyahi ho,
zamin bakir khani,*

*sara jungal dahi ho,
to kaun dega pani?*

Old Mother Hubbard
went to the cupboard
to get her poor dog a bone;
when she got there,
the cupboard was bare,
and so the poor dog got none.

*Dharmi Dai
handi tak gayi
kutte ko dene har;
wahan jab ayi
to kuchh na payi;
rah gaya rozahdar!*

Humpty Dumpty sat on a wall,
Humpty Dumpty had a great fall.
Not all the Queen's horses,
    not all the King's men
could put Humpty Dumpty
    together again.

*Hamti Damti charhgaya chat;
Hamti Damti girgaya phat,*

Raja ka paltan, Rani ke ghore

Hamti Damti kabhi na jore.

Little Bo-peep
has lost her sheep,
and doesn't know where to find
    them;
let 'em alone

*Chhoti Momeri
hargaya bheri,*

*kidhar se moh gayin gum;
chhuti rahenge*

and they'll come home
and bring their tails behind them!

Goosey, Goosey, Gander,
where shall I wander?
Upstairs, or downstairs,
or in my lady's chamber?
Old Daddy long-legs
wouldn't say his prayers;
take him by the left leg
and throw him downstairs!

Dickory, dickory, dock;
the mouse ran up the clock,
the clock struck one
and down she run,
Dickory, dickory, dock.

Little Miss Muffet
sat on a tuffet,
eating her curds and whey,
when a great ugly spider
came and sat down beside her
and frightened Miss Muffet away.

Higgledy, piggledy, my fat hen!
She laid eggs for gentlemen,
sometimes eight, and sometimes
    ten,
higgledy, piggledy, my fat hen!

Mary, Mary, quite contrary,
how does your garden grow?
With silver bells and cockle shells,

and pretty maids all in a row.

to ghar men awenge,
wa sab ke pichhe dum!

Hans, Hans, Raj Hans,
kidhar jane hota?
Upar jawen, niche jawen,
Bibi-ji ka kotha?
Budha Behuda
chhor diya namaz;
gor dharke phenk de,
pir pae-daraz.

Dekho re, dekho re, dekh!
Ghari bajegi ek!
Jab ghanta hua,
to kud para chuha,
Dekho re, dekho re, dekh!

Mafiti Mai,
dalai malai
ghas men baithke khai,
jab bara sa makra
uski sari ko pakara,
bhage Mafiti Mai!

Hakali, makali, murghi mera!
Anda pare barah terah;

par ke bhej de sahib ka derah;
hakali, makali, murghi mera!

Miriam meri tirchni terhi
phuta gulistan?
Chandi ka ghanta wa kauri
ka panta
wa larki khub jawan.

T.F. Bignold, ICS, *Leviora*, 1863–88

Sanskrit Quatrains from A.G. Shirreff, ICS,
The Tale of Florentius, 1914

I

Yours the loss if your dispraises
put the virtuous to shame;
turn a torch down, still it blazes,
but your fingers feel the flame.

X

Vicious men will seek to show
vice like theirs in men of worth.
After eating much, the crow
wipes his bill on the clean earth.

XV

If a man is bad at soul,
effort will not put him right:
you may wash and scrub a coal,
but you will not make it white.

XXIV

As a pot retains impressions
once upon the wet clay traced,
so the earliest moral lessons
will not ever be effaced.

8

Some Portraits

From Tales of a Tahsildar, 1950

Introduction

The name and office of *Tahsildar* will be familiar to all those who have served in northern India. For others, a brief description will have to suffice.

A Tahsildar was, and, so far as I know, still is a subordinate official in charge of a revenue and civil area of administration called a *tahsil*. In the United Provinces (now Uttar Pradesh), where the scene of this story is set, a tahsil varied in size from four hundred to six hundred square miles, with a rural population of anything from one hundred and fifty thousand to three hundred thousand, living in some one hundred and fifty to three hundred villages. There were, on average, four tahsils to a district, six districts to a division, and eight divisions or forty-eight districts in the province. For those who find it easier to visualize these areas in terms of the size of an English county, an average district in the UP, was from about one-third to one-half the size of all three ridings of Yorkshire combined.

The British adopted the tahsil from the Moguls, who, in turn, inherited the unit, if not the name, from the system of village and district administration which they found in existence when they started to devise their own system of administration in India.

It would not be a great exaggeration to say that the whole revenue and civil administration of the British Raj in northern India hinged on the Tahsildar—the man who actually saw that the order was carried out. To give some idea of the variety of tasks required of him, he had, for instance, to measure the rainfall on the roof of the tahsil building, the political temper of a dozen of the largest villages in his area, the damage done to crops by severe frost, storm or flood; he might have to arrange for the extermination of locusts, a duck shoot for the

Governor, or for fuel and hay for a brigade on their way through his tahsil.

It was sometimes best not to enquire too closely through how many mouths the order passed before it reached someone so humble that he could pass it no further and so had to carry it out himself. The important thing was that the job was done and that everyone, including the Tahsildar himself, knew that he was the man responsible. I doubt if things have changed much in this respect since the British left the country.

Of course, there were incompetent and dishonest Tahsildars. But, on the whole, they were a worthy and resourceful body of men, the nuts and bolts of the so-called steel frame.

It gives me great pleasure, therefore, to pay tribute in this tale to the memory of a most remarkable Tahsildar, who served under me in the district of, shall we call it, Mustypore, in the early nineteen-thirties. He was a political refugee from Afghanistan, one of several who were given sanctuary and employment in British India. He was a square peg in a round hole if ever there was one; but by sheer strength of character he had got right on top of his job with the minimum of exertion, to the exasperation of his more orthodox and hard-working contemporaries.

Without more ado, therefore, let me introduce you to Rahat Ali Shah, Tahsildar of Islamabad, and the story of:

The Failed Racehorse

One day in the hot weather of Mustypore was very like another; one was grateful, therefore, for any break in the monotony.

One morning, about 10 a.m., while I was in my office seeing visitors, I heard a commotion in the drive outside my bungalow, punctuated by stentorian imprecations. I recognized the voice of my friend, Sardar Rahat Ali Shah. Shortly afterwards the orderly came into my office and said:

'The *Tahsildar* Sahib of Islamabad—am I to show him in?'

He spoke with due deference, but I sensed the antipathy in his voice, conveyed with all the subtlety and resignation of the East. I pretended not to have noticed the meaningful inflexion and merely nodded. The orderly went out; as he passed through the doorway, letting the *chick* fall behind him, he sighed, loudly enough for both me and my visitor to hear, but in such a way that he could explain it away, if necessary, as due to the flatulence to which he was a chronic victim.

It was Saturday. On Wednesdays and Saturdays official visitors had precedence. On other days of the week, if they wanted to call during visiting hours, they had, ordinarily, to make an appointment by letter or telephone and in any case to convince the Collector that their business was urgent. It was past 10 a.m. now; a number of officials had already called during the cool of the morning. The orderly reflected that the visiting hours were nearly over . . . and there was still a certain contractor . . . but he would have to go and put him off now.

The oriental visitors' room or, as often as not, verandah, provides an incomparable school of psychology, infinitely superior, in this respect, to our own impersonal if more impartial arrangements. In the long lazy hours the *peon*, reclining against the wall of the cool verandah, seldom wastes his time merely minding his own business. He is the direct medium between visitor and visited, interpreting the one to the other with all the embellishments of imaginative artistry; applying to one the respectful innuendo, to another the soft answer, to a third the contemptuous aside. To read the mind of every visitor like an open book, to know the likes and dislikes of his employer and to calculate to a nicety his reactions and their value to an impatient suitor—this was a fascinating study and a profitable business.

When Rahat Ali Shah entered the room it appeared, suddenly, to be full, but at the same time brighter. He was a vast man with a heavy but soldierly figure, a big head, prominent beaked nose, full high-coloured cheeks and fierce pointed and waxed mustachios. He carried himself in public with a natural and somewhat aggressive swagger, reserving for informal moments and for his friends a disarming ingenuousness and a boisterous sense of humour of the schoolboy variety. His huge gusts of laughter exposed devastating visions of stained and broken teeth as inescapable as the infection of his mood. He was very shrewd and incurably lazy.

The dress he was wearing now was of his own devising—his 'revenue-collecting' he called it. Above elegant pointed light yellow

---

Chick door-screen made of split bamboo   *peon* orderly, messenger

boots he wore red leather gaiters the colour of raw liver, tight khaki twill breeches and coat with enormous side pockets. At his hip was a service revolver, a .45 Colt, and across his swelling chest a leather bandolier containing three cartridges, all of different bores and none of them fitting his weapon. I had asked him about these once; he had replied with a grin that they impressed the ignorant and deceived the knowing, and snapping open the Colt, he had shown me that it was fully loaded save for the one empty chamber at the top. In his belt was stuffed a heavy riding crop; he twisted in his hands a vulgar check tweed cap of the style known as 'Apache'. It was appallingly greasy.

Some might smile in private but none, even of the local notables, would have dared to take any further liberty at the expense of Sardar Rahat Ali Shah, Afghan political refugee by the will of Allah and Tahsildar of Islamabad by the grace of the British Government.

As he sat down we both smiled, as it might be at the opening of a school boxing match. I had not summoned Rahat Ali and he was not in the habit of calling frequently and unnecessarily over trivialities like some officials. The odds were, therefore, that he had come to ask some special favour.

We talked 'shop' for some time—a necessary but tiresome preliminary. The Sardar, I knew, considered it distasteful for gentlemen to have to refer to such matters at all. That these formal discussions sometimes included items which proved awkward to explain did not, however, unduly depress him; nor did the fact that promotion to the rank of Deputy Collector did not come his way, was not likely to come his way and was no longer a subject to which he referred. He was satisfied if his superiors appreciated a gentleman, something that no amount of promotion could create. He had an efficient assistant and led, on the whole, an easy life, clouded only at intervals by the prospect of the civil lock-up, for the shadow of his debts, like that of his figure, never grew less. He was an incorrigible gambler.

After a while the conversation turned—I do not recall why—to gardening. The Sardar brightened; he knew something about gardens. I mentioned a project for turning a disused camping ground near his tahsil into a fruit and vegetable garden. He listened politely and, at the proper moment, reeled off a catalogue of the salads he would grow there, ending with cucumber and a loud smack of his lip to show his appreciation of that succulent plant. I remember expressed some doubts:

'I generally find it pretty tasteless in this country,' I said.

The Sardar considered the matter and then said, 'Sir, the growing of cucumber is not so easy—I mean the Kabuli cucumber, not the watery 'loki' of Hindostan which, as you say, is tasteless. First the seed must be obtained from Kabul and, if possible, a little of my native soil . . . but I weary your honour with unnecessary details.'

There was a pause and then he said, speaking more slowly, 'Talking of my native soil reminds me of the object of my visit. I would request your honour to grant me three months leave from April. I wish to prosecute my claim to our ancestral lands,' and he looked me straight in the eye as he always did when he lied most unblushingly.

I went to the cabinet and got out his file.

'Here,' I said, pulling a letter out of the file, 'is the last and final word on your ancestral lands; you have seen it already. Mr White did his best for you—it would be a sheer waste of time and money to pursue the matter further.'

Rahat Ali did not even bother to look at the letter.

'Sahib,' said he, 'we have a proverb in our own country: "There are many ways of answering an awkward letter but only one way of getting rid of an importunate beggar." Besides, the leave is due to me.'

There was more than a hint in his eyes, though they were smiling. I thought it over. I ought to refuse, of course, or, at least, to say I would think it over. But I smelt a story; if I refused there would be no story. And all the time those sharp smiling eyes were boring into me I made up my mind to risk it. Things were quiet and his assistant well able to look after the tahsil in his absence.

'Very well,' I said, 'I will grant your leave—on condition that you keep out of debt and out of Afghanistan.'

'Sahib, the word of an Afghan Prince. . . .'

'Is not worth one midday pull of a *punkah* coolie,' I finished the sentence, perhaps rather rudely.

Rahat Ali looked hurt, but nevertheless grinned broadly, revealing some of his appalling dentures.

The interview was over. I walked out into the verandah with my visitor. There was a scraping of chairs and some rather too elaborate salutations, hand to brow. I noticed a fat contractor, whose name, I remembered, had already been announced three times by the

---

Punkah large swinging cloth fan, pulled by a cord or leather thong

orderly—that, of course, explained the sigh; a seedy-looking Anglo-Indian whose trousers were so short that they did not even cover the top of his boots and a veiled woman with three very dirty children. I would have to see them all, even the contractor, though I was tempted to put him off in the hope that he would demand his consideration back from the orderly who was still sulking. But first, as I knew from experience, there must take place the Sardar's ceremonious leave-taking.

In the drive stood the most fearsome-looking horse I had ever seen. In the first place it was too long in the body, but let that pass—the same could be said of many a good horse. The Indians have a saying for it: 'The moon has struck him.' A quick appraisal of its neck told a tale of unbending obstinacy; it stood over at the knees and one hock was enlarged, permanently I had no doubt. The shoulders were a bit better, though too narrow. But it was the head and, in the head, the mouth which claimed foremost attention. It was the ugliest head and the hardest mouth I had ever seen. The bit matched the beast—a wicked affair covered with spikes. The saddle and harness were definitely provincial. The animal was dozing on three legs, a thin brown liquid drooling occasionally from one corner of its mouth. A small boy, who looked suspiciously like one of the club tennis boys, stood at its head with a self-conscious smirk on his face.

The Sardar was watching me closely. When I turned to him he broke into one of his most disarming smiles, 'My new charger,' he explained proudly, 'a failed racehorse, BSc'

He looked anxiously at me to see if I would play the game.

'BSc?' I queried obediently.

'Bombay Steeple-chaser, your Honour,'—Rahat Ali stood to attention and saluted. We both laughed and I looked again at the animal incredulously.

'I can prove it, your Honour—I have his pedigree somewhere,' and the Sardar started to search his voluminous pockets. Watching him with amusement, I saw his expression change; some awkward afterthought must have presented itself. It had: he confessed to me later that the pedigree was in a letter of recent date addressed to him at the Taj Mahal Hotel. He had never obtained leave to go to Bombay. The search diminished in tempo and died. Rahat Ali shrugged his shoulders.

The Most Fearsome–Looking Horse

'I cannot find it, but no matter; my wife's uncle bought him in Bombay after his career had finished.'

'And gave him to you, I suppose?'

'Yes, your Honour, a family arrangement. He is a long-distance stayer, slightly unsound, but a horse of great heart.'

'He looks stronger in the head than in the heart. But it is better, sometimes, as the saying goes, not to look a gift horse too closely in the mouth,' I said rather fatuously.

Rahat Ali bowed slightly from the waist—I felt I had deserved the gesture.

'And have you brought him in specially to show me, Rahat?'

'Your Honour: I also had work on the way.'

'Do you mean to say you have ridden him all the way this morning?'

'Your Honour, we walked.' I calculated swiftly.

'Twenty-seven miles in a day! I shall be losing a valuable officer; anyway, mind you walk him home as well.'

'I will try my best to follow your Honour's advice but the decision does not rest with me entirely.'

I must have looked incredulous, for he added earnestly,

'Sir, he raced under the name "Lord Nelson"—there is still a touch of the old devil left in him.'

Rahat Ali grinned and, putting on his Apache cap back to front with the peak pulled well down behind, saluted smartly once more and walked down the steps. The boy brought the horse over. Its seventeen hands towered above the Tahsildar: I was curious to see how he would mount.

There was quite an impressive gallery now—all my visitors, the orderlies, the *mali* and his small son with the basket of weeds still on his head; my wife's *darzi* removed his big toe from the handle of the sewing machine just in time to catch the spectacles which fell off his nose as he rose to watch the fun. Rahat Ali was enjoying himself. The steps were ornate, and flanked by a wall in the shape of a curling scroll. The Sardar mounted the tail of the scroll and from there, with surprising agility, transferred himself suddenly onto the correct section of equine anatomy before him. Ostentatiously he pulled out a rupee and, bending down, gave it to the small boy. I caught the words 'Telegraph office—at once'. It was only then that I noticed that

---

Mali gardener   *darzi* tailor

the boy had a telegram form in one hand, already written out. As the Sardar straightened up he deliberately caught my eye and grinned,

'Just to tell my lawyer my leave is sanctioned,' he said and, after saluting me with his crop, he brought it down smartly across the relief map that was his charger's quarters. They moved off unhurriedly.

'Nothing,' I thought, 'will ever induce that animal to break out of a walk.'

But I was wrong. Apparently the trouble started on the Mall, only three hundred yards from my gate. As they passed a stationary refuse cart (locally known as an 'iron-clad'), its lid wedged open with a stick, the racehorse snorted. It would be nice to believe and better still to be able to prove that the smell offended his well-bred sensitivity. It is, I fear, more likely that, in common with most horses, he disliked buffaloes. The buffalo standing within the shafts of the ironclad shied and the lid of the cart fell with a monstrous clang. The buffalo bolted but only for a few yards; the racehorse also bolted but did not stop till it reached the tahsil, thirteen and a half miles away and 45 minutes later, in circumstances of which I myself was a witness.

But, before that, the whole affair nearly came to an abrupt and fatal end at the main Cantonment cross-roads. The traffic constable was passing one of those large canvas frames on wheels which advertised the current programme at the local cinema. Hearing the thunder of flying hooves bearing down on them, the two men in charge abandoned the frame right across the road down which the runaway was approaching at what the Tahsildar afterwards assured me was thirty-eight miles an hour, for he had his stop-watch out to 'test his uncle-in-law's veracity' as he put it.

Rahat Ali just had time to note that the film was called 'The Lover's Leap' before they rose at the obstacle and sailed through the top half, tearing the lovers apart and overturning the frame. At the canal bridge, where he should have turned off down the canal road, the most direct route, Rahat Ali shouted at the *tendal* standing by the gate. The man did not hear what he said but took in the situation and very sensibly rang up both the tahsil and my bungalow which he knew the Tahsildar had been visiting. His message was brief and clear: 'Tahsildar Sahib bolted on horse'.

Tendal head of a gang of labourers, mainly associated with the maintenance of canals

I had seen the seedy Anglo-Indian in boots and was just about to see the fat contractor but got out my car at once. I was a little anxious for Rahat Ali because of that harness. At the canal bridge the tendal told me the runaway had not gone by the canal road so I followed the district-board road in the hope of picking up some clues. Two miles out of the Cantonment I came on a *dhobi* still contemplating with despair the washing which he had spread out to dry along the verge of the road. A trail of dusty hoof-marks betrayed the passage of the fugitive. Five miles down the road I found the Apache cap and, two miles further on, fragments of harness—unidentifiable—which might have been there for twenty minutes or twenty centuries.

Here the district-board road crossed the canal again and took a wide turn before eventually reaching the tahsil, whereas the canal road ran straight there. I was confident that Lord Nelson would make for his home port and only hoped he was sensible enough to realize that this was no longer Bombay. I was no longer really anxious about Rahat Ali; if he had not fallen off up to this point the odds were that he was still abroad. But having come so far I was curious to know how the affair would end. The canal road was gated. After eleven miles I doubted whether a retired steeplechaser would face the gate, especially as there was an easier way cross-country, the normal approach to the tahsil for anyone on horseback. I had the key so opened the gate onto the canal road and 'stepped on it'. I was just in time to see the finish.

There was quite a good field out, thanks to the tendal's timely message. The doctor was there with a stretcher, the litigants had left the court with their lawyers, all the clerks and orderlies. There was the usual miscellany of touts, beggars, *sadhus*, small boys and pi-dogs.

As I approached the tahsil I saw, some three fields away, a great roll of yellow dust, billowing across the plain. It thinned, and then dramatically parted to reveal the iron horse making very heavy weather of the fallow. In fact he was approaching at a slow trot which grew slower every minute—and no wonder. The saddle was empty but, slumped over the withers, hung the Tahsildar, fast as a leech, his arms round the horse's neck, his whole sixteen stone bearing through his knees on the unfortunate animal's forehand. 'Bolted on horse' was a very apt description. Gradually the disjointed trot dropped to a

---

Dhobi washerman   *sadhu* ascetic

Lord Nelson at Full Gallop

broken stumble. Half a field away and I could distinguish the set jaw
and clenched teeth of the rider, his hair, eyebrows and mustachios
white with dust, his cheek pressed close against the lathered neck of
his mount.

With the last agonized stumble the charger reached the tahsil
compound and sank to its knees. Rahat Ali slid forward gently onto
his. For a moment horse and rider rested immobile, exhausted. The
doctor rushed up with his stretcher: the Sardar released his vice-like
grip and waved him away. Stiffly he disentangled himself and surveyed his distressed mount. There was a look of the utmost satisfaction on his face. A *syce* appeared unobtrusively, the horse heaved
laboriously to its feet and was led off, dead lame.

'Water,' croaked the Sardar hoarsely; there was a rush to comply.
When it came he swilled out his mouth generously and blew his nose.
Then he turned to his audience. For the first time he noticed me; he
drew himself up and saluted. Then he closed one eye and gave a
monstrous wink with the other,

'Your Honour, the Half-Nelson Touch!'

R.V. Vernède, ICS, 1950

A Nightmare Durbar

(An official endeavouring to obtain from Simla some information on
Durbar dress was informed by wire: 'Trousers are not to be worn at
Delhi')

Throughout the day they echo in my head
and through the night from sunset to reveille
these nightmare orders haunt my sleepless bed—
'Trousers must not be worn at Delhi'

I strive to solve the riddle, but in vain,
as they who sought the whereabouts of Kelly,
the meaning slips elusive through my brain—
'Trousers must not be worn at Delhi'

And as the ribald, shameless message fled
from Cutch to Bude, from Quetta to Clovelly,
the British Empire turned a deeper red—
'Trousers must not be worn at Delhi'

Syce groom

I see a row of lean official shanks,
dyspeptic colonels trembling like a jelly,
Panjandra shivering in Jaeger ranks—
'Trousers must not be worn at Delhi'

Heads of Departments pass with foreheads bent
and blushing 'pine for that which is not' (Shelley),
'Trousers must not be worn at Delhi'

And grief is writ upon my consort's brow,
Hinc illae lacrimae!—it's hard on Nelly,
She'd been invited, bought her kit, and now—
'Trousers must not be worn at Delhi'

> J.M. Symns, Indian Educational Service, 1913

'*Hamlet*' at a Bengal Fair

It was at an up-country fair in Bengal that we saw 'Hamlet' played by a native company, and it rounded off our fairing in an instructive and delightful way. We had gone to the fair—the Collector and his wife and two babes, Clothilde and I—because the Collector had been asked to open it and the rest of us wanted to go. We travelled by means of one *tonga*, four ponies, and two elephants, one of the elephants acting as perambulator when the tonga got stuck at particularly bad bits of the road. We did the forty miles in two days, which is good travelling for Bengal, especially as we got a leopard on the road. Speaking exactly, the leopard was off the road about three hundred yards, in a grass jungle. A little cloud of vultures circling over it, waiting for it to finish its meal, gave us the clue to its whereabouts. It was wounded by the first bullet, and made a spring for Clothilde's legs, Clothilde being on the pad of our second elephant, but it missed its spring, and the next shot finished it.

Apart from the leopard, the dust was the most noticeable thing on the road, especially as we drew near to the fair in the afternoon of the second day. If there had been any wind we should have been buried by the dust. Two hundred acres of sandy sun-baked plain crowded with street after street of booths, alive with a hundred thousand natives, and countless elephants, camels, cattle and ponies. That was

---

Tonga (tanga) a light two-wheeled carriage, drawn usually by ponies, holds two passengers behind; space for luggage is in front beside the driver

the fair, and the whole air tingled with the dust of it, and we gulped it down red-hot from the sun as we rode in. Doctor Johnson never drank at a sitting more tea than I did when we arrived at the dak bungalow.

From its verandah there was plenty of fair life to be seen without stirring. Bhutanese, sturdy pig-tailed buccaneers, rode past driving before them a herd of their shaggy little ponies—the sort Bengali sub-inspectors of police love to acquire and ride, partly because they have superbly flowing manes and tails, partly because they can be cantered twenty miles without stopping, under an Indian sun. These ponies, like Nicholas Nickleby at Dotheboys Hall, are remarkable for their straight legs. The ordinary Bengali *tat*, ridden or burdened from its cradle, never has straight legs, and an Englishman told me of one he had borrowed for the day whose legs were set at such weird angles that it could not stand up until he got on its back. Then his weight pressed them in the directions necessary for balance, and it went with spirit after dacoits. After the Bhutanese, and swallowing their dust, would go bullock-carts bringing merchants' wares, the drivers walking; then perhaps the merchant himself, magnificent on a tat going *cuddam*, bath slippers on feet that nearly touched the ground, and no stirrups. It is a curious pace, this cuddam, and I do not know if it obtains outside India. The pony using it seems to flicker or shiver along, and there is no more motion for its rider than for a lady in a bath-chair. It is eminently suited for the *babu*, being both slow and comfortable, and I take it that the nearest English equivalent to it was the amble of the monks of Chaucer's time on their way to Canterbury.

Then a north-country man would go by on a camel, and some local *zemindar* would trot his native devil-eared horse past us as fast as it would go, in the hope that we were watching and admiring. We did watch for a time, and afterwards Clothilde and I set out for the fair. The formal opening was to be next day, but we wanted to see it by ourselves first, and without ceremony. The desire was a vain one. Almost before we had passed the gate leading in, we were sighted by a policeman, who either wished to earn merit or to assert a brief authority. At any rate, he constituted himself our vanguard, and after

---

Dak bungalow rest-house for travellers   *tat* (*tattoo*) native-bred pony   *cuddam* amble of a horse   *babu* person of distinction, often used disparagingly   *zemindar* (*zamindar*) landowner

that, peace and privacy were impossible. Authority in this country—where, according to the babu, liberty calls loudly to the soul of every man—is not regarded as a means to an end. It is an end in itself and a veritable passion. If a Bengali sees a chance of bullying, he will take it, and his fellows will accept the part of victims with almost equal ardour. Our way through the fair, crowded though it was, was clear enough, since we only wanted to stroll along examining the booths at our leisure. But the policeman would not have it so. To left or to right he would dart, shoving some poor unfortunate who might conceivably have been in our way had we been going that way. The person shoved would seek credit by shoving the man nearest to him, who would shove the next, who would shove a boy, who would shove a smaller boy. Nobody seemed to mind. Indeed they all seemed to enjoy it except ourselves, who wanted peace instead of this hurly-burly, and could not command the policeman in his native tongue. We were rescued by coming across Mr Chundar.

I had met Mr Chundar once before. He was a middle-aged Bengali babu, engaged as estate agent and general factotum to the Raja upon whose grounds the fair was held. Under Mr Chundar's aegis the fair took shape, and he was responsible for its success or failure. But his chief glory was that he was a Barrister-at-law of—let us say at random—the Middle Temple. Barristers-at-law in this country enjoy a certain dignity and distinction. Mr Chundar also enjoyed what dignity a solar topi and a frock-coat and trousers might give him. But it was some years since he had trod the Middle Temple, and I suppose he had forgotten that with a frock-coat one used not to wear in the Middle Temple an old pair of white canvas shoes with the laces unfastened.

All the same, we were grateful for the appearance of Mr Chundar at that point in the fair, for he spoke English, and though he did not sympathize with us, and appeared to be a little shocked by our desire for peace and privacy, he did, when I insisted, rid us of the policeman. Left to ourselves we went up and down the booths. It was essentially a country fair—a fair for the *ryot*—and though there were some local industries represented, 'Made in Birmingham' or 'Made in Germany' stared at one from most of the stuffs and wares. Not so with the animals, of course. Neither Frankfurt nor Birmingham can produce

---

Solar (*sola*) topi light weight hat made from the dried pith of a plant *ryot* (*raiyat*) farmer or cultivator

live elephants or camels, and the ponies were all native. Fairly good elephants were to be had for about two thousand rupees. The camels were poor and thin. The keen northerners had not brought of their best to this southron market. We saw more of these animals on the following day, for after the opening ceremony, we were escorted to a small circular racecourse, set in the middle of the fair, to witness some camel races. When I say 'we', I mean the Collector and friends and Raja and suite. We took our seats on a set of drawing room furniture upholstered in green brocaded satin, which had been brought from the Raja's house and placed ready for us under a canopy. A local band was also ready for us and struck up 'God save the King', as soon as we appeared. The tune was sonorously rendered, but the bandsmen had not that *esprit de corps* that some conductors insist on, and several of the musicians wandered into other tunes that may have been more beautiful but did not tone in.

Perturbed, perhaps by the music, the first contingent of camels, four in number, refused to start. Their riders did their best, and the barrister-at-law, in his white tennis shoes, addressed them at some length, at first imperiously and then with tears in his eyes, but the camels would not budge and had to be withdrawn. Graceful conversation by the Raja carried us over this little hitch and the second line of camels was brought forward. Again the band struck up and again the camels exhibited a puritanical objection to racing. The barrister-at-law became frantic; he skipped in his tennis shoes and waved his arms commandingly. His efforts were useless. The bandsmen, entranced by this struggle of wits between the babu and the beasts, strayed into all sorts of keys and tunes, some of them forgetting to play altogether. Suddenly, three of the camels started. For some ten yards they ran a neck-and-neck race; then two of them hit their shins against the hurdles between which they were racing, and collapsed like a pack of cards. There is nothing that goes down so dramatically and so completely as a camel. The third creature was made of sterner stuff. Annoyed by being compelled to start and enraged by the strains of the band, the brute, without stopping, turned its head right round and made maddened efforts to eat its rider. It was an interesting sight, the unfortunate rider slipping farther and farther back to escape that long snarling neck, the camel galloping *ventre à terre*, with its head serpentined round and its nose and lips all mixed up in a spitting, biting fury. It was better than a race; it was a duel and we watched fascinated. Would the camel complete the circle without devouring its rider, or would the latter, by deft tugging, bring it to a stop? The unexpected happened. Rider and camel both being taken up by their internecine strife, forgot that their course lay between hurdles, and in the midst of a peculiarly vicious snap lost their direction and knocked a hurdle down. For a moment the camel paused, startled by the noise and the presence of the excited onlookers. Then perceiving directly in front of it the Raja and ourselves—a strange and offensive group—it came straight at us, screaming with passion. With remarkable presence of mind we all rose at once and placed the drawing room suite between ourselves and the infuriated beast.

Another six paces and it would be on us. The band had ceased to play, the crowd hummed with suppressed horror. In the distance I saw the barrister-at-law awaiting with horror-struck eyes and clasped hands the inevitable catastrophe. Then, with a superhuman effort, the rider gave a last tug at the rope-bridle, and the camel fell in folds before us.

'I think he ought to get the prize,' said the Collector's wife to the Raja, as we reseated ourselves with all the dignity possible under the circumstances. The Raja smiled courteously and said that the camel was an animal uncertain to ride, but useful, especially in the north. Still, he cast a menacing look at Mr Chundar when that barrister-at-law came up to regret the unfortunate issue of the camel race, and to consult his Honour as to whether this camel, as having kept its feet longest, was to be adjudged victor, or whether it should be disqualified, as having maliciously made for his Honour's party with intent to damage. The Collector's wife decided sportingly in favour of the fighting camel, and the band seized this moment to give us 'God save the King' again. To restore us we had tea and cake of the wedding pattern handed round, and after that we inspected the prize-winning cattle. The prize cow gave two and a half quarts of milk and the second gave two and their prizeworthiness was not wholly apparent to the naked eye. Prizes to encourage the cattle industry of a district are excellent things, but it seems even these may be put to wrong uses. My friend the planter told me of a zemindar in a district he had once known, whose tenants were always the winners of the rupees he offered as prizes. The reason they won was that they could be made to give the rupees back more promptly and easily than outsiders. I hoped that this was cynicism, or at least a solitary example of the misuse of prize-money at cattle-shows. No doubt there is a temptation in India to appear charitable without being so. There is a temptation in all countries, but India has its particular variety. How? It is a vast place, with many landholders in it, all filled with an aimiable desire to distinguish themselves. In order to become distinguished under the British Raj, it is well to assist the commonweal in some form. Charity, such as is involved in the offering of prize-money for cattle-breeding, is a simple and straightforward form of assisting the commonweal. But suppose that you are but a poor man, though a landholder. Why, then be charitable still, but drop the straightforwardness. It is just as simple not to be straightforward. Give the prize-money as before, but see that you get it back again. If that is too extreme a thing to do, and it is, there are many other ways in which charity in India works out a little less simply than it is supposed to do. You will perhaps meet a rich zemindar who puts down his name for a large donation to some well-advertised and well advertising public work, and forgets to forward the cheque when called upon. Another will send a generous yearly subscription—for the first year only.

I seem to be wandering from the fair, and the chief event in it, which was the performance of 'Hamlet'. It took place later in the day, beginning at eight o'clock and lasting until after midnight. It was a command performance, to which the Raja had invited us, and it was therefore not to be witnessed without due ceremonies. We had 'God save the King' as we entered and were ushered by the barrister-at-law to the drawing-room suite in green brocaded satin from which we had been privileged to watch the camel fight. It was now the front row of the stalls in the big marquee that constituted the theatre. We had 'God save the King' about two minutes later, when the Raja and party entered, and it may be said on this subject generally that if repetition of this tune can be taken as the best assurance of loyalty, nobody in Bengal need have the least doubt of its prevalence there.

Hardly were we all seated when Clothilde and I, being less experienced than the rest, leapt from our seats as a bomb exploded one pace from us, followed by two more in swift succession. They were, of course, only salutes—tributes to our combined importance—but they left me somewhat deaf for the rest of the evening. I cannot say I was sorry for this, because of the orchestra. The orchestra was composed of two players. It was not the band of the morning that had made the camels so restive. That band was somewhere outside and was only used when 'God save the King' was required. The inside orchestra consisted of (1) a harmonium-player on the left wing of the stage; (2) a tom-tom player on the right. For many minutes that evening these two monopolized our attention. The harmonium-player was a young slim Bengali in a coat and *dhoti*, patent-leather shoes, and what used to be called in England a polo cap—a brown pork-pie shaped cap set jauntily on one side of his head. His action on the harmonium was inimitably careless and graceful. No European master, I venture to think, has ever expressed such contemptuous mastery over his instrument. He would play it with one hand daringly, as a novice rides a bicycle to show off to a friend, while with the other he fetched betel from his waistband and transferred it to his mouth; or he would, in an ecstasy of abandonment, crash both fists on to the harmonium, crossing the keyboard and coming back again before one could stiffen one's muscles to bear it. I have not heard a musician like him either before or since. I am not skilled in music, nor do I know the Indian notation. But one is accustomed to regard the harmonium as a sober instrument. Conceive it in all its long-drawn reverberating fulness attacked by something as wild as a jungle-cat, as heavy as a jungle bear, and you have some idea of the excruciating sounds which that young man in the polo cap extracted from it. Compared with him the tom-tom player—a square person, who sat on a small kitchen table, with his bare feet protruding into the stalls—was a soothing nonentity. When he played his loudest, which he often did, he only slightly subdued the nerve-stretching ululations of the harmonium-player. Moreover, they rarely combined or got on to their stroke together. I could not make out the rules, but I fancy they played when they felt like it. When the harmonium was too intent on betel-chewing to play up, the tom-tom droned away for a few minutes. When the harmonium, refreshed by the leaf, dashed himself at the most discordant notes he could find, the tom-tom took a breathing space. Sometimes, like two omnibus drivers moved to rivalry, they raced one another on their respective instruments, but there was never any question as to which won. The tom-tom was distinctly second fiddle.

What—it may be asked—had this orchestra to do with 'Hamlet'? What—as far as that goes—has any orchestra to do with 'Hamlet'? As a matter of fact, this pair was pretty busily engaged, for 'Hamlet' in Bengali is—if I may attempt a definition—a musical tragedy of imbroglio. Whenever the action palled (and there was lots of action)

Dhoti loin-cloth, traditionally worn by male Hindus

one of the players sang a song—not so much accompanied by the orchestra as defied by it. Hamlet himself was the only man that had a chance against the harmonium, and that was due to the penetratingly nasal quality of his voice. Again, I have never heard anyone so nasal as Hamlet. He reminded me sometimes of a Swiss yodeler heard nearby; sometimes of a Venetian boatman singing 'Funicoli-funicola' on the water outside one's window. He never reminded me of Hamlet.

Here, before I enlarge upon the acting, I will set down, act by act, the programme of the play, of which the plot was specially printed for us in English, so that we might understand. 'The Plot in Short' it is called. It lies before me as I write. I give it as printed:

The first scene opens with the King chatting with the Queen in a room in the castle. He then feels drousy and subsequently falls asleep; whereupon the Queen sends for her husband's brother, Farrukh and induces him to drop poison in his ear. The King dies of its effects, and the Queen gives out, importunately attributing the cause to a serpent's bite. Jehangir mourns his father's death and Akhtar, his friend and associate, comforts him.

This, it will be observed, is Shakespeare, though not in the order we know it. Liberties have been taken, but what actors have not taken them? The point to be noticed is that the plot serves India admirably. Look at the Queen importunately attributing her husband's death to a serpent's bite. It is thoroughly Bengali. Official returns of today attribute an enormous proportion of deaths among natives to snake-bite; individuals say that the variety of snake is a human one. Anyhow the pit understands. Jehangir is, of course, Hamlet. In his make-up he conformed to the English tradition so far as to wear Hamlet's black cloak. Otherwise he was an innovator: he wore rowing shorts, puttees and a pair of football boots; also a big pistol in his girdle, such as highwaymen used to carry and, fully exposed like a decoration, a large gun-metal watch and chain over his heart. We supposed at first from its calibre that the watch was merely a decoration, but this was not the case. It had a dramatic value too. You remember the famous lines in Act III:

Tis now the very witching time of night,
when churchyards yawn and hell itself breathes out
contagion to this world: now could I drink hot blood,
and do such bitter business as the day
would quake to look on.

Well, Hamlet wanted to make quite sure that it was the very witching time of night when he could drink hot blood, and he consulted the gun-metal watch accordingly. There was a pleasing accuracy about this that seems to indicate that the actor took the view that Hamlet's madness was only feigned.

What with the watch and the pistol, Hamlet's was a sporting rather than a historical make-up, and I think Akhtar (Horatio) was rather envious of it. He was somewhat of a Job's comforter, but nothing was likely to quell Hamlet's mourning. In Bengal it had to be of a pristine ceremonial order. There was no possible doubt about its intensity. He simply 'waked' his father and, with the assistance of the harmonium, approached the banshee at its best. One felt that some action was bound to ensue, and Act II was, in the circumstances, a little disappointing. Here is the syllabus of it:

At the opening of the second act, Farrukh in court putting on the guise of anxiety for Jehangir's safety shows concerns and inquiries. Mansoor the Wazirzada falls in love with Meharbano. Suleiman enters and a conversation passes on. Akhtar recounts the accident of the grave to Suleiman. Seeing Jehangir entering, Suleiman withdraws. Akhtar questions Jehangir who confides him with the disclosure. Mansoor in frenzy declares his love for her.

'Her' is, of course, Meharbano or Ophelia. The chief interest of this Act consisted in the introduction of the characters new to Shakespeare. Mansoor the Wazirzada was the most important in rank, but Suleiman was more important dramatically. He was, so to speak, the Shakespearian clown Indianized. Later he became the first grave digger. The thing about him was that he was a black man, not a brown one. That was the comedy of him. The audience laughed when they saw him. Everything he said was a joke. I could not make out quite what his relations were with the other characters, but I do not think they greatly mattered. The clown may enter anywhere. He gives relief and in this Act one was grateful for relief. The acting was all very emotionally pronounced, and the harmonium was at his most energetic.

With next scene we come to a room where the Queen is seen merrymaking with Farrukh. Then enters Humayun, the Lord Chamberlain who, soon after, is despatched to console the Prince. The Queen then gives publicity to her union with Farrukh. Meanwhile the Wazir tries to solace the Prince who hears him with flightiness and cynical disdain, and pours forth in soliloqy his horror at his mother's marriage.

'Soliloqy' hardly expresses the prolonged and rampant vocalism to which Hamlet, undeterred by the harmonium and the tom-tom, treated us. But here again, of course, his horror had to be very great. Not only was his Queen Mother marrying her husband's murderer, but she was remarrying; and to a Hindu Hamlet a widow's marriage would justify any outburst. The Queen's action represented shamelessness and passion or was supposed to; but none of the women in the play showed any emotion comparable with that of the men. It would not have been proper, or, presumably, like real life. This took away a good deal of the interest of Ophelia, who had her chance in Act IV as the programme shows:

Meharbano's giving went to her love for Jehangir. Her maids-of-honour soothing her. Jehangir's going to his father's grave. Akhtar's and Suleiman's oversighting him. The opening of the grave. The appearance of the Ghost and informing him of his death.

Meharbano gave, it seemed to me, the very meekest possible 'went' to her love for Jehangir, and her maids-of-honour had little or no difficulty in soothing her, though they spread their consolations over a considerable period. Meharbano was a small artiste, with the voice of a field-mouse. She had on a cherry-coloured satin dress, which reached barely to her knees, and—with a view to captivating Hamlet, no doubt—a pair of European black stockings. No shoes. The exceedingly loose fit of the stockings led to an unintentional piece of by-play at one point. She was giving 'went' to her love by squeezing a tiny pocket-handkerchief, of which she made a good deal of use throughout, passing it through her fingers and laying it on her breast, when she accidentally dropped it. In Bengal, when you drop a thing, there is no bothering to stoop and pick it up. You use your foot. One of the courtiers—not very courteously—nudged Ophelia and pointed to the fallen handkerchief. Absent-mindedly she put out one big toe at it gracefully, half raised it, and then had the mortification of seeing it fall again. She had forgotten her stockings, prisoned in which her prehensile toe had lost its cunning. She had to bend down to get it. If this Act gave Ophelia her opportunity, it also gave Hamlet his—at the graveside. That was after the appearance of the Ghost, who looked, it must be allowed, more English than the rest of the dramatis personae, and had a fairly good speaking part. Roused by his tale, Hamlet did a sword dance, preparatory to taking vengeance. It was

Hamlet—'Soliloqy'

a great effort, that dance, lasting roughly for ten minutes, Hamlet doing Indian Clubs with his sword and shrieking at the top of his voice throughout. The young man at the harmonium appeared to be really moved by it and, as it were, challenged Hamlet to musical combat. The conclusion was a foregone one. Hamlet did his best, and it was a good best, but a man cannot contend with a harmonium indefinitely. The young instrumentalist reduced Hamlet to a hoarse impotence at the end, and went on by himself for a minute or two, just to show what an agony of organ notes the harmonium can give forth when the Master wills it. After this Act V, though full of incident, seemed in its way quiet. The following events took place:

Mansoor's and Sahelin's jesting with each other in the way. His going in the garden with their help. Declaring of his love for Meharbano. Her declining. Coming of Jehangir and his killing Mansoor. Coming of every one in the Tamasgarh. Farrukh and Jehangir witnessing performance. The death of all.

The programme is not perfectly clear. As far as I remember, it was Mansoor who got into the garden 'with their help'. Anyhow, he was in the garden, and Jehangir came and killed him. He killed him by coming up behind and shooting him in the back with a shiny new rook rifle. Someone must have given Jehangir the rook rifle at the end of Act IV, perhaps instead of a bouquet. I feel sure he had not possessed it before, or he would have brought it on. The wound produced by it, besides being mortal, was of a very painful nature, and Mansoor depicted it with consummate skill. Indeed, apart from Hamlet's sword dance and the death of all which followed later, there was nothing more appreciated by the audience. On the English stage deaths are for the most part swift, if dramatic. In Italian Opera they take longer very often, but the efforts of the artistes are concentrated rather on getting their notes out successfully than upon depicting the postures and writhings in unduly harrowing last throes. Singers are too careful of themselves and, as a Mansoor had set a sublime example, and all, when death came upon them, strove to equal his performance. I do not know why the death of all occurred, but it did so quite suddenly— I should say it began to do so quite suddenly—and, though it came in the form of the poison-cup, pistol shots and the stab of a dagger, it came with similar lingering, writhing, hair-raising preliminaries. Ophelia retained her breath the longest, and there was in her end a distinct touch of the star actress. She had stabbed herself in good time with a very large stage dagger wrought of wood and silver paper which puckered, but she reserved her death to the last. She allowed about a quarter of an hour for the others to writhe, and then staggered to the front and was about to fall. A difficulty presented itself. The stage was so packed with the dead bodies that space adequate for the decease of the heroine was lacking, at any rate in the front. Ophelia showed the practical common-sense that has before now distinguished artistes. Nothing daunted by the affair of the handkerchief, she again used her foot to kick one of the crowd in the ribs. With one of those convulsive spasms that have been known to occur even after death, he jerked himself to one side. Hamlet was the other too forward corpse, but a poke in the back enabled him to perform the same phenomenon. Then Ophelia could really abandon herself to die, and did so.

There was sustained applause from the whole theatre, particularly from the front row of the stalls and, after it was over, Mr Chundar who had been busy between the Acts handing us chocolates and biscuits, came up to find out what we thought of the performance.

'You like it? You think it was well acted?' he asked us smiling, but with an anxious eye on the Raja at the same time.

We all declared that we liked it immensely, and that it had been acted very finely indeed, and Mr Chundar's smile expanded and expanded. Only the Raja had yet to speak and he, judging that we had been pleased and satisfied, and that none of the failure attaching to the camel race could be assigned to this performance, said very graciously, 'Yes, it was well acted. You shall tell the company that they did well.' And he added courteously to me, who sat on one side of him, 'It is a good little play. Yes.'

Next moment the band outside struck up 'God save the King' for positively the last time, and to these loyal strains we walked out into the Bengal night. It was a lovely night. The stars glittered from a black velvet sky, and in the starlight, as we drove back, we could see the shrouded Bengalis shuffling home along the dusty road. Though we had all of us been seeing Shakespeare's 'Hamlet', I had the strange feeling that we were moving in some time and place that were pre-Shakespearian.

R.E. Vernéde, *An Ignorant in India*, 1911

Kuppan the Beggar

Kuppan the old beggar sits under the tamarind tree
with the sweetmeat man and the loafers, hard by the town bridge end.
Naked and ribbed and wrinkled, the crows are his company,
the sparrows call him brother, and the mynas know him a friend.
He sits in the drift of the shadows and the traffic's come and go,
and he cries: 'Aiyo, 'S'amil oh, Maharaj Aiyo!'

The carts come in from the country, the fishers come up from the sea;
constable, clerk and coolie, they stop in the shade and chat.
He hears the news they are telling, he sees what there is to see,
cross-legged there by the roadside, rocking himself on his mat.
He cries upon heaven for mercy, he wails to mankind in woe,
but—there's little happens in this town he doesn't contrive to know.

He never worked and he never will, but he knows he can always live;
it's sometimes brinjal or chillies, it's sometimes *ragi* or *dhal*,
but it's always the scraps and the oddments this one and that one give.
and Kuppan blandly impartial, drones a blessing for all.
Nothing to own and nothing to lose, he has stripped Fate's armoury
    bare,
for he knows what will happen to-morrow—he knows and he doesn't
    care.

He knows he'll be there to-morrow, next week, next month, next year;
nothing to plan or think of, never a worry or doubt,
and always the people passing and always something to hear.
He sees day in with the sunrise, he sees late evening out—
free, gratis, for nothing, with a front seat in the show,
and he cries: 'Aiyo, S'amil oh, Maharaj Aiyo!'

Charles Hilton-Brown, ICS

No. XVI: The Civil Surgeon

Throw physic to the dogs, I'll none of it.

Perhaps you would hardly guess from his appearance and ways that
he was a surgeon and a medicine-man. He certainly does not smell
of lavender and peppermint, or display fine and curious linen, or tread
softly like a cat. Contrariwise.

He smells of tobacco, and wears flannel underclothing. His step is heavy. He is a gross, big cow-buffalo sort of man, with a tangled growth of beard. His ranting voice and loud familiar manner amount to an outrage. He laughs like a camel, with deep bubbling noises. Thick corduroy breeches and gaiters swaddle his shapeless legs, and he rides a coarse-bred Waler mare.

I pray the gods that he may never be required to operate upon my eyes, or intestines, or any other delicate organ—that he may never be required to trephine my skull, or remove the roof of my mouth. Of course he is a very good fellow. He walks straight into your drawing-room with a pipe in his mouth, bellowing out your name. No servant announces his arrival. He tramples in and crushes himself into a chair, without removing his hat, or performing any other high ceremonial. He has been riding in the sun, and is in a state of profuse perspiration; you will have to bring him round with the national beverage of Anglo-India, a brandy and soda.

Now he will enter upon your case. 'Well, you're looking very blooming; what the devil is the matter with you? Eh? Eh? Want a trip to the hills? Eh? Eh? How is the bay pony? Eh? Have you seen Smith's new filly? Eh?'

This is very cheerful and reassuring if you are a healthy man with some large conspicuous disease—a broken rib, cholera, or toothache; but if you are a fine delicately-made man, pregnant with poetry as the egg of the nightingale is pregnant with music, and throbbing with an exquisite nervous sensibility, perhaps languishing under some vague and occult disease, of which you are only conscious in moments of intense introspection, this mode of approaching the diagnosis is apt to give your system a shock.

Otherwise it may be bracing, like the inclement north wind. But, speaking for myself, it has proved most ruinous and disastrous. Since I have known the Doctor my constitution has broken up. I am a wreck. There is hardly a single drug in the whole pharmacopoeia that I can take with any pleasure, and I have entirely lost sight of a most interesting and curious complaint.

You see, dear Vanity, that I don't mince matters. I take our Doctor as I find him, rough and allopathic; but I'm sure he might be improved in the course of two or three generations. We may leave this, however, to Nature and the Army Medical Department. Reform is not my business. I have no proposals to offer that will accelerate the progress of the Doctor towards a higher type.

Happily his surgical and medical functions claim only a portion of his time. He is in charge of the district gaol, a large and comfortable retreat for criminals. Here he is admirable. To some eight or nine hundred murderers, robbers, and inferior delinquents he plays the part of maitre d'hotel with infinite success. In the whole countryside you will not find a community so well bathed, dressed, exercised, fed and lodged as that over which the Doctor presides. You observe on every face a quiet Quakerish air of contentment. Every inmate of the gaol seems to think that he has now found a haven of rest.

If the sea-horse on the ocean
own no dear domestic cave,
yet he slumbers without motion
on the still and halcyon wave;
if on rainy days the loafer
gamble when he cannot roam,
the police will help him so far
as to find him here a home.

This is indeed a quiet refuge for world-wearied men; a sanctuary undisturbed by the fears of the weak or the passions of the strong. The poor burglar burdened with unsaleable 'grab' and the reproaches of a venal world sorrowfully seeks an asylum here.

Look at this prisoner slumbering peacefully beside his huqqa under the suggestive bottle tree (there is something touching in his selection of the shade of a bottle tree: Horace clearly had no bottle tree; or he would never have lain under a strawberry (and cream) tree). You can see that he has been softly nurtured. What a sleek sturdy fellow he is! He is a covenanted servant here, having passed an examination in gang robbery, accompanied by violence and prevarication. He cannot be discharged under a long term of years. Uncovenanted pilferers, in for a week, regard him with respect and envy. And certainly his lot is enviable; he has no cares, no anxieties. Famine and the depreciation of silver are nothing to him. Rain or sunshine, he lives in plenty.

His days are spent in an innocent round of duties, relieved by sleep and contemplation of τὸ ὄv. In the long heats of summer he whiles away the time with carpet-making; between the showers of autumn he digs, like our first parents, in the Doctor's garden; and in winter, as there is no billiard table, he takes a turn on the treadmill with his mates. Perhaps, as he does so, he recites Charles Lamb's Pindaric Ode:

Great Mill!
that by thy motion proper
(no thanks to wind or sail, or toiling rill)
grinding that stubborn-corn, the human will,
turn'st out men's consciences,
that were begrimed before, as clean and sweet
as flour from purest wheat,
into thy hopper.

Yet sometimes a murmur rises like a summer zephyr even from the soft of luxury and ease. Even the hardened criminal, dandled on the knee of a patriarchal Government, will sometimes complain and try to give the Doctor trouble. But the Doctor has a specific—a brief incantation that allays every species of inflammatory discontent. 'Look here, my man! If I hear any more of this infernal nonsense, I'll turn you out of the gaol neck and crop.' This is a threat that never fails to produce the desired effect. To be expelled from gaol and driven, like Cain, into the rude and wicked world, a wanderer, an outcast— this would indeed be a cruel ban. Before such a presentiment the well-ordered mind of the criminal recoils with horror.

The Civil Surgeon is also a rain doctor, and takes charge of the Imperial gauge. If a pint more or a pint less than usual falls, he at once telegraphs this priceless gossip to the Press Commissioner, Oracle Grotto, Delphi, Elysium. This is one of our precautions to guard against famine. Mr Caird is the other.

G.R. Alberigh-Mackay ('Sir Alibaba K.C.B.'),
Twenty-One Days in India, 22 November 1879

9
Nostalgia

Night in Camp, U.P.

And so to bed. And so to sleep? Perhaps.
Not while the camp is getting under weigh,
the carts assemble, creaking as they sway;
men talk and cough; bells tinkle, a dog yaps.
Soon I shall hear the forward tent collapse
with a dull thud, and then its whole array
of tent-pegs must be malleted away
to a tattoo of syncopated taps.
No sleep for me while these alarums last;
no sleep for me,—yet why do I complain?
In the not distant future I forecast
that often I shall long to hear again
these voices of the night and long in vain,
when camping is a pleasure of the past.

> A.G. Shirreff, ICS, in Oswald Couldry,
> *Sonnets of East and West*, 1951

In Spring Time

My garden blazes brightly with the rose-bush and the peach,
and the *koil* sings above it, in the *siris* by the well,
from the creeper-covered trellis comes the squirrel's chattering
    speech.
But the rose has lost its fragrance, and the koil's note is strange;
I am sick of endless sunshine, sick of blossom-burdened bough.
Give me back the leafless woodlands where the winds of
    Springtime range!
Give me back one day in England, for it's Spring in England now!

Koil (koel) Indian cuckoo   *siris* a common Indian tree

Through the pines the gusts are booming, o'er the brown fields
    blowing chill,
from the furrow of the ploughshare streams the fragrance of the loam,
and the hawk nests on the cliff-side and the jackdaw on the hill,
and my heart is back in England 'mid the sights and sounds of Home.

Rudyard Kipling, *Departmental Ditties*, 1886
(last four lines omitted)

The Moon of other Days

Beneath the deep verandah's shade,
    when bats begin to fly,
I sit me down and watch, alas!
    another evening die.
Blood-red beyond the sere ferash
    she rises through the haze;
sainted Diana! can that be
    the Moon of other days!

Ah! shade of little Kitty Smith
    sweet Saint of Kensington!
Say, was it ever thus at Home
    the Moon of August shone,
when arm in arm we wandered long
    through Putney's evening haze,
and Hammersmith was Heaven beneath
    the Moon of other days?

But Wandle's stream is Sutlej now,
    and Putney's evening haze
the dust that half a hundred kine
    before my window raise.
Unkempt, unclean, athwart the mist
    the seething city looms,
in place of Putney's golden gorse
    the sickly babul blooms.

Glare down, old Hecate, through the dust,
    and bid the pie-dog yell,
draw from the drain its typhoid germ,
    from each bazaar its smell;

yea, suck the fever from the tank
  and sap my strength therewith;
thank heaven, you show a smiling face
  to little Kitty Smith.

         Ibid.

The Land of Regrets

Yea, they thought scorn of that pleasant land.

      Psalm 106, v. 24

What far-reaching Nemesis steered him
from his home by the cool of the sea?
when he left the fair country that reared him,
when he left her, his Mother, for thee,
that restless disconsolate worker,
who strains now in vain at thy nets,
O sultry and sombre Noverca!
  O Land of Regrets!

What lured him to life in the tropic?
Did he venture for fame or for pelf?
Did he seek a career philanthropic,
or only to better himself?
But whate'er the temptation that brought him,
whether piety, dullness or debts,
he is thine for a price, thou hast bought him,
  O Land of Regrets.

He did list to the voice of the Siren,
he was caught by the clinking of gold,
and the slow toil of Europe seemed tiring
and the grey of his fatherland cold;
he must haste to the gardens of Circe;
what ails him, the slave, that he frets
in thy service? O Lady sans merci!
  O Land of Regrets.

From the East came the breath of its odours
and its heat melted soft in the haze,
while he dimly descried thy pagodas,
O Cybele, ancient of days;

heard the hum of thy mystic processions,
the echo of myriads who cry
and the wail of thy vain intercessions
through the bare empty vault of the sky.

Did he read of the lore of thy sages?
of thy worship by mountain and flood?
Did he muse o'er thy annals?—the pages
are blotted with treason and blood;
thy chiefs and thy dynasties reckon?
the armies—he saw them come forth
o'er the wide stony wolds of the Dekhan,
o'er the cities and plains of the North.
He was touched with the tales of our glory,
he was stirred by the clash and the jar
of the nations who kill *con amore*,
the fury of races at war.

'Mid the crumbling of royalties rotting,
each cursed by a knave or a fool,
where Kings and fanatics are plotting,
he dreamt of a power and a rule;
hath he come now, in season, to know thee;
hath he seen what a stranger forgets,
all the graveyards of exiles below thee,
        O Land of Regrets?

Has he learnt how thy honours are rated?
has he cast his accounts in thy school?
With the sweets of authority sated,
would he give up his throne to be cool?
Doth he curse Oriental romancing,
and wish he had toiled all his day
at the Bar, or the Banks, or financing,
and got damned in a common place way?

Thou hast racked him with duns and diseases,
and he lies, as thy scorching winds blow,
recollecting old England's sea breezes
on his back in a lone bungalow;
at the slow coming darkness repining,

how he girds at the sun till it sets,
as he marks the long shadows declining
    o'er the Land of Regrets.

Let him cry as thy blue devils seize him,
O stepmother, careless of fate;
he may strive from thy bonds to release him,
thou hast passed him his sentence—'Too Late';
he has found what a blunder his youth is,
his prime what a struggle, and yet
has to learn of old age what the truth is
    in the Land of Regrets.

                  Sir Alfred Lyall

The Successful Competitor No. I

Oh! for the palmy days, the days of old!
when writers revelled in barbaric gold;
when each auspicious smile secured a gem
from Merchant's store or Raja's diadem;
when 'neath the pankha frill the Court reclined,
when 'Almah wrote and Judges only signed;
or, lordlier still, beneath a virgin space
inscribed their names and hied them to the chase!
Chained to the desk, the worn Civilian now
clears his parched throat and wipes his weary brow.

The Successful Competitor No. II

Oh honoured Yule! I would I were like thee,
dispensing justice 'neath a sheltering tree;
and, guided less by training than by tact,
could pounce unerring on the trail of fact;
for in those days—'tis long ago, my friend—
Law was the means, and Justice was the end;
now Rhadamanthus revels in a flaw,
and weaks injustice while he teaches law.

. . . .

For the good Magistrate, our rulers say,
decides all night, investigates all day;
the crack Collector, man of equal might,
reports all day and corresponds all night.

T.F. Bignold, ICS, 1871

IX: *Ballade of Myself*

My great great grandfather, so goes
    the cycle of a century,
was first of all my line who chose
    the service of John Company.
    The generations pass, and I,
the last of all his lineage,
    once more his ancient calling ply;
I have a goodly heritage.

From where her tropic ocean flows
    by groves of fadeless greenery,
to where her everlasting snows
    flush faint across the sapphire sky,
    beyond the scope of memory,
this land has been the spacious stage
    of all my race's history;
I have a goodly heritage.

Well that my lot is cast with those
    whose hearts have heard the calling high
to guide her aims, to quell her foes,
    to realize her destiny,
    her name has been their battle-cry,
her welfare a sufficient wage
    for which to live, for which to die:
I have a goodly heritage.

Mataram bande! Thus the tie
that links us lasts from age to age
    in ever fuller sanctity.
I have a goodly heritage.

A.G. Shirreff, ICS, 1918

Mataram bande! Hail, motherland!

The Wooden Foil

For me no place upon this earth
has quite the same romantic worth
     as that small station
in which my salad days were spent
'Out East' and when I underwent
     my first Probation.

Still through the mists of far off years
distinct as yesterday appears
     (to me *nirvana*!)
my compound with its white-washed wall,
the old bazaar, the dusty mall,
     the cool gymkhana.

Once more beneath its roof I see
the Judge, the gallant DSP,
     the Colonel's daughter,
and recollect with what an air
she used to sit her chestnut mare;
     how sweet I thought her!

Once more within the bar I meet
the Doctor in his favourite seat
     (a trifle fatter?)
hobnobbing with the Padre and
discussing loudly, glass in hand,
     some burning matter.

The two Miss Browns again display
the modes of some forgotten day
     that knew not Mammon;
angelic friends—I see you yet
intent on your beloved picquet
     or else backgammon.

Your special province, the bazaar,
entailed more strenuous work by far
     than many men did;

Nirvana state of infinite bliss

your faith wrought miracles, 'twas said;
at least the destitute were fed,
    the sick attended.

One tenth of what I have to-day
I counted then most liberal pay,
    and always spent it;
the panther and the jinking boar
took most, but if I wanted more—
    why—someone lent it.

We'd no electric fans or ice,
or polo at a fancy price,
    or nine hole courses;
and though, I think, we must have heard
of motor cars, we much preferred
    to ride our horses.

We'd friends in many a village then,
salt of the earth they were, though men
    of humble showing,
the men to whom this troubled land
now seldom cares to understand
    the debt that's owing.

O days of peace (and pride no less)
when Cleon and his noisy press
    had no effect, or
as ever since the world began,
some people chose to trust a man
    like my Collector.

Such spacious days are over now,
and times have changed—well anyhow
    I go to-morrow.
Mine is the wooden foil; and yet
I take it not without regret
    and heartfelt sorrow.

            'Momos'

Dak Bungalows

'Dak bungalows?' said Jobson, 'Dak bungalows!' said he:
'I hit the trail for the Gorgeous East in 1893,
and I've wandered thirty years from Chatrapur to Comorin
till there's not a rest-house on the road but I've cast anchor in.
Dak bungalows! Don't speak of them to me.

You get 'em walled with native mud and roofed with rotten thatch,
where the bats hang up in hundreds and the baby cobras hatch;
you get 'em built of Government brick, topped off with country tiles,
where the rains come through and weep on you and the sun looks in
    and smiles;
but there's precious few that you could call a catch.

You get 'em cocked up on a hill, or settled by the Bay,
or buried in some jungle glade where the bamboos creak and sway.
You get 'em miles from anywhere, or plumb in the bazaar
where the pi-dogs howl and the drums beat up and the native
    voices jar.
You get 'em different day by weary day.

You get 'em new and type-design with the doors and windows plumb,
with the varnish sticky on the chairs and a smell of paint and gum.
And all along the Grand Trunk Road you get 'em old and odd
with tombstones in the compound and an air of 'Ichabod';
yes, you've simply got to take them as they come.

By railway, river and canal, by road and bullock track
you do your stage and stay your night and write your name and pack.
Come hill or dale, come hot or cold, come lightning, thunder, rain,
you wander on from house to house and wander out again,
and know you're never likely to be back.

You get 'em good, you get 'em bad, you get 'em worse or worst,
but always there's a likeness odd betwixt the last and first,
because of those who all these years from house to house have plied
and either joked and jollified, or ached and wept and died.
And the ghosts come back that blessed the place—or cursed

Dak bungalow   rest-house for travellers

Dak Bungalow

And often on a stuffy night, 'mid the pi-dogs' maddening din,
when the breeze is dead and it's ninety-odd and the pochies bite
    like sin,
I read the names in the bungalow book and wonder why and how
they came along and where they went and what they're doing now;
and I sometimes see their faces looking in.

Familiar faces—Magistrate, Policeman, Engineer,
Forest and Salt and all the rest who humped their battered gear
from Bungalow to Bungalow and paid twelve annas rent
and spread their kit and packed again and did their work and went,
but somehow left behind an atmosphere.

Dak bungalows! A funny life—arrive, unpack and flit.
But my name is in a hundred books and you may learn from it
that a certain friendless fellow—one Jobson, it appears—
made his home in these houses for a spell of thirty years,
and groused a lot—but, Oh! he's loth to quit.'

                      C. Hilton-Brown, ICS, *Both Sides of Suez*

To an Up-Country Bus

Time was—and not in very distant ages—
the proud Pro-Consul travelled in a cart
and moved by slow and meditative stages;
nor knew this rush and its attendant rages
    so hard upon the heart.

Nor rabid yet with restlessness and crazy
with that mad itch to speed the parting hour,
serene he passed from morn to twilight hazy.
And life, dear life, so liberal to the lazy,
    unfolded like a flower.

And days turned over with the sober rustle
of leaves in some digested drowsy tome—
those days ere Bharata was bound in bustle;
those happy days ere Hind had heard of hustle,
    and haste was left at home.

The madmen came, and with them came disaster,
like that sad Queen whom Alice, gentle girl,
met on her travels, shrieking, 'Faster! Faster!'
They seized on this unhurried land and cast her
    into the common whirl.

Along their rails rude locomotives thundered,
filling the land with fret and fume and fuss;
wires wedded those whom kindly space had sundered
and last to this lone Arcady they blundered
    and ran a motor-bus.

And so you came with roar and rust and rattle
down the rough road the stately bullocks trod,
when we went herded not in trucks like cattle,
and life was life and not a raging battle,
    and speed was not yet God.

Headlong you came in murderous intrusion,
roaring aloud that peace must give you place.
Symbol of progress—that insane illusion;
King of the creed whose motto is confusion,
    Prime Minister of Pace.

And I? Yes I, despite these maledictions,
must mount once more upon your hateful back
and taste anew your pains and your restrictions,
your stony blows, your triturating frictions,
    more ruthless than the rack.

I, who in earlier ages, ere the virus
of maniac haste had poisoned human wit,
might have gone forth by cart, content as Cyrus,
a ten-mile stage (or shorter if desirous)
    and loved each yard of it,

now—that confusion's cup and folly's flagon
be drained completely—now must sit and stew
for forty miles—O soul-destroying dragon!
O loathsome Juggernaut! O weary wagon!
    O beastly bus!—in you.

                          Ibid.

Garhwal

From tiger-land forest to high Himachal,
from bare terraced foothills to blue Bireh Tal;
as deep as your valleys, as strong as your forts,
your memory lingers, Garhwal, in my thoughts.

Remember those forest camps in the Terai,
the smell of log fires, the peacock's harsh cry,
the call of the *kakar* not far from the kill?
O heart stop your beating! Remember the thrill?

The tree-shaded spring by which, resting his load,
the coolie relaxed on the sun-blistered road;
the jingle of mule-trains, the smell of old dung,
the crazy rope-bridges which dipped as they swung?

The pipes of the shepherds, the pulse of the drums,
the gay wizened pensioners 'toting' their rums;
the shy little maidens with mountainous loads,
the ash-bedaubed holy men squatting like toads?

Vestigial terraces whorled like a thumb,
vertiginous path above precipice plumb;
the peasants' respect for invisible powers,
the Trident of *Siva*, the Valley of Flowers?

High camps above valleys of indigo blue
in tents hanging ghostly and silvered with dew;
sky colours which *Lakshmi* herself might have worn,
hot ashes of sunset, the oriole dawn?

The Flame of the Forest outburning the day,
the primulas' glory, the jasmine's bouquet;
the air of the mountains as heady as wine,
the incense of deodar, perfume of pine?

The troops of grey *langurs* who crashed through the trees
like a squall swooping down over mountainous seas;
the shrilling of marmots, the bark of the fox,
the baying of sheep-dogs protecting the flocks?

Kakar the barking deer found in Terai and Bhaber forests below the Himalayan foothills   *Siva* Hindu God of destruction   *Lakshmi* Hindu goddess of beauty, wealth and good fortune

The little stone temple beside the high pass,
the mica-strewn bridle road shining like glass;
the mountains, cloud-riding, rose-capped in the glow,
the red rhododendrons against the white snow?

Though oceans divide us and memories fade,
I still see your mountains in glory arrayed;
though Time that old robber has torn us apart,
your magic, Garhwal, is secure—in my heart.

R.V. Vernède, Oxford 1950

Back East

The whirling months go round
and back I come again
to the baked and blistered ground,
and the dust-encumbered plain,
and the bare hot weather trees,
and the Trunk Road's aching white;
Oh, land of little ease!
Oh, land of strange delight!

Home's woods October-red,
Home's pastures summer-green
are as a memory fled,
are as they ne'er had been;
ragged our gardens stand
in destitute undress,
stripped by the sun-god's hand,
his all too close caress.

Eyes that the other day
on ordered scenes could feast,
now meet the disarray,
the turmoil of the East—
rough-hewn, unfinished stone,
mad landscapes half begun,
sketched out in monotone
and slaughtered by the sun.

Yet oh! the sailing moon
the tropic night becalms:
and oh! the lit lagoon
beyond the coco-palms;
kind evening's ecstasies,
and the deep dome of night;
Oh, land of little ease!
Oh, land of strange delight!

   Madras Mail, 1933, published in
C. Hilton-Brown, ICS, *The Sahibs*

Far Away

Beyond the hills and far away
I know a small Headquarter station,
where three hours work at most per day
commands official approbation.

No 'sense of grave injustice' galls
the spleen of local politicians;
no suppliant throng at noon-time calls,
or sends preposterous petitions.

No rival parties disagree;
no Council there prolongs its sittings;
and every bungalow is free,
with marble floors and bathroom fittings.

Long annual vacations cure
the strain of regular employment,
and special rates of pay ensure
a maximum of leave's enjoyment.

No fever there or sudden chills
the varied joys of living veto;
no sleepless wretch at midnight kills
the hovering sandfly or mosquito.

Since wit and wisdom both abound,
all nuisances are soon abated,
and obvious bores are either drowned,
or quietly defenestrated.

No gossips have been known to thrive,
no snobs have ever lightly boasted,
and humbugs there are skinned alive,
and alcoholic cooks are roasted.

They live like Gods throughout the year,
dull care magnificently scorning;
they drink immeasurable beer,
and sleep till ten o'clock each morning.

They spend their days in cultured ease
and utterly despise conventions;
they think exactly as they please,
and thus secure quite early pensions.

                       'Momos'

The Old 'Koi-Hai'

An antique air surrounds his chair
in drawing room and club,
an old *Koi-Hai*, left high and dry,
a far away look in his eye,
and dreaming of the days gone by,
the good old days in 'Jub'.

The leather tan proclaims a man
whose world has lost its hub:
no ready thralls to answer calls,
no boy brings whisky when he bawls,
stretched out at ease in marble halls,
as once in good old 'Jub'.

No butler staid in gold brocade
to serve him with his grub;
no *khidmatgar* to bring cigar,

Koi hai anybody around! universal expression used by the British when calling
for servants to take orders for drinks at the club or a hotel   Jub Jubbalpore, in
Central Provinces   *khidmatgar* waiter

or fill the brown tobacco jar;
no chauffeur now to wash his car,
no sweeperess to scrub.

No *bhisti* thin with glistening skin
to fill his morning tub;
no *dhobi* foots to iron his suits,
no beaters to attend his shoots,
no bearer to remove his boots,
or give his back a rub.

No fellow bore to share the floor,
no crony at the pub;
no boy around to feed the hound,
no *syce* to bring the pony round;
no-one to meet on common ground
and no-one left to snub.

When all is said—he's not yet dead,
he goes on paying his 'sub';
so please be kind and do not mind
if he continues to remind
us of the days he's left behind—
the good old days in 'Jub'.

R.V. Vernède, ICS, 1950

To Mrs Heber

If thou wert by my side, my love!
How fast would evening fall
In green Bengala's palmy grove,
Listening the nightingale!

If thou, my love! wert by my side,
My babies at my knee,
How gaily would our pinnace glide
O'er *Gunga*'s mimic sea.

Bhisti water carrier   *dhobi* washerman boy manservant   *syce* (*sais*) groom
Ganga the river Ganges

I miss thee at the dawning grey
When, on our deck reclined,
In careless ease my limbs I lay
And woo the cooler wind.

I miss thee when by Gunga's stream
My twilight steps I guide,
But most beneath the lamp's pale beam
I miss thee from my side.

I spread my books, my pencil try,
The lingering noon to cheer,
But miss thy kind approving eye,
Thy most attentive ear.

But when of morn and eve the star
Beholds me on my knee,
I feel, though thou art distant far,
Thy prayers ascend for me.

Then on! then on! where duty leads,
My course be onward still,
O'er broad Hindostan's sultry mead,
O'er bleak Almorah's hill.

That course, nor Delhi's kingly gates,
Nor wild Malwah detain,
For sweet the bliss us both awaits
By yonder western main.

Thy towers, Bombay, gleam bright, they say,
Across the bright blue sea,
But ne'er were hearts so light and gay
As then shall meet in thee!

The Right Rev. Reginald Heber D.D.,
Lord Bishop of Calcutta 1783-1825

(With acknowledgements to Theodore Oliver Douglas Dunn, Author of *Poets of John Company*, 1921, Bodleian Library, 2805 d 78.)

The English Flag

Winds of the World, give answer!
They are whimpering to and fro—
and what should they know of England
who only England know?

. . . .

Never was isle so little,
never was sea so lone,
but over the scud and the palm-trees
an English flag was flown.

. . . .

Never the lotos closes,
never the wild-fowl wake,
but a soul goes out on the East Wind
that died for England's sake—
man or woman or suckling,
mother or bride or maid,
because on the bones of the English
the English Flag is stayed.

. . . .

The dead dumb fog hath wrapped it—
the frozen dews have kissed—
the naked stars have seen it,
a fellow-star in the mist.
What is the Flag of England?
Ye have but my breath to dare,
ye have but my waves to conquer.
Go forth, for it is there!

Kipling, *Barrack Room Ballads and
Other Verses*, 1892

Index of Authors and Their Works